WHAT IS THE DIMENSION OF TIME?

Time, Mass and Energy: The Hows and Whys

Dr. Mitchell Albert Wick

authorHOUSE®

AuthorHouse™
1663 Liberty Drive
Bloomington, IN 47403
www.authorhouse.com
Phone: 1 (800) 839-8640

Published by AuthorHouse 01/04/2018

ISBN: 978-1-5246-2257-2 (sc)
ISBN: 978-1-5246-2256-5 (e)

Print information available on the last page.

- Equation of Everything

Spherical Space-time expanding as a current into fluid-like space-time

space-time=space/ mass

The outward push of Dark Energy yields 5x10^39 radians of reciprocal curvature or flattening

The Cosmologic Constant flattens space-time by curving it outward as a push by equation $\Lambda=1/R2$ where R=space-time curvature

Equation of

Time with reference to Gravity and Inertia. R(t)=[+or - **Rg** ab+Λ*ba*⊗*iℏ*Λ *ba ρ ab*

R(t)=+ or –R g abΛ+(*COSMOLOGIC CONSTANT*)/R ab]_WHAT IS THE

DIMENSION OF TIME? by Dr . Mitchell Albert Wick

FORWARD AND INTRODUCTION

In this author's previous works ,a one inch equation was shown to explain all the phenomena of nature .In this boo k this equation was utilized to show how many equations there are in nature. In addition, it was shown why the expansion of space-time and the 750 billion galaxies in our universe is accelerating geometrically and why the Critical Density of the Universe (point when expansion slows ,stops and becomes a contraction)will never be reached. The Cosmologic Constant is show to be related intimate with Dark Energy which pushes the galaxies away from each other while the pull of space-time in the same Cartesian Coordinates is acting as a "super jet-stream effect "where space-time which acts like a Perfect Fluid(a known fact)has a current which is being pulled by other space-time traveling at an almost im-mentionable velocity outward. The reason being this pull is explained in depth in this book.

The total curvature of space-time was explained in the previous book "Mega-physics II:An Explanation of Nature" as 5×10^{-29} radians where 2(pi)radians is 360 degrees but in this book space-times curvature is explained in depth as to when it curves inward(gravity) and when it curves outward(anti-gravity or the effect of the Cosmologic Constant and Dark Energy).

Once the mechanism of the Perfect Fluid of space-time is understood, one can understand the mechanism of how time changes space and why space-time is a continuum(nothing doesn't exist and never did or will exist)

There is a mechanism to the origin of the string dimensions and macroscopic dimensions which formed the Multi-verse;not just our Universe ;and how there were an infinite number of totally parallel non-intersecting planes prior to the formation of the dimensions of strings(more than 26).There is an explanation of how a perturbation caused an oscillating time paradox where time's arrow(forward)and time's arrow(backward)caused an oscillating of time constricting then dilating these infinite parallel planes causing a centrifuge effect where the string dimensions and macroscopic dimensions formed and from there the Multi-verse. This author is certain that anyone with curiosity about answering these questions will be captivated by this book and hopefully discuss it with professors and researchers.

TABLE OF CONTENTS---

---------CHAPTER ONE: DOES TIME EXIST?

The dimension of time is described in the line element as ds^2=dx^2+dy^2+dz^2-c^2dt^2+dr^2. Time was always thought of as the sequencing of events as if all events occurred simultaneously there would be absolute chaos without sequential **ordering..**
For example, time is required for us to exist after "The Big Bang" and for the earth to exist after "The Big Bang" as without it we would not exist for the development of carbon based intelligent life such as "us" and would violate the Anthropic Principle. We would be theoretically occupying the same space as a temperature of 2.74 degrees kelvin as the galaxies were being formed and the same space as when expansion has approached maximum resulting in "Heat Death".

It has been illustrated that "A Big Crunch" may cause time's arrow to reverse causing time to revert towards zero without reaching it. Using the equation of everything and Einstein's Equation for Relativistic Gravity it has been determined by this author that the dimension of time (T)=+ or -1/2 R g a b /R a b where R g a b is the space-time curvature metric of the metric g a b and is the effect of gravity and R a b is the Ricci Tensor describing inertia of the mass that's doing the curving .In other words time is directly proportional to gravity and inversely proportional to inertial mass where the constant k~ 1/c^2or exactly as the energy density of matter .If you consider time as R(t) or R^ d with reference to space R a b c as Riemannian or Lorenzian(curved) Space-time R a b c d and R a b c as Euclidian Space or Minkowski(flat) space-time R a b c d(where R stands for "region" in topologic space in which the tensor acts) where the time component is R^ d or R d then R ^d=R(t)=-1/2R g a b or +1/2R g a b/R ab. Of course the Bianchi Identity indicates that R a b c=-R c b a=R^ c b a=R^a b c for flat Minkowski space-time with R^ d=R d=-R d= -R d in the case where time's arrow moves forward and where time's arrow moves backward in a "Big Crunch" R^(-d)=-R (-d)=+R d.

CHAPTER TWO - DOES TIME'S ARROW REVERSE IN A BIG CRUNCH?

In the case of a Big Crunch space-time constricts toward an infinite curvature point angle=n (2 *pi*) radians where n approaches infinite curvature the dimensions which were 26 compactified(curled up and "compacted") dimensions reduced to 10 in string theory or 11 with super-gravity would have the curling up or compactification less important making the hidden dimensions more important as a "Big Crunch" is approached. As the dimensions have been demonstrated in my previous book "Mega-physics II: An Explanation of Nature" were to approach infinity without reaching it as each second of arc in a sphere of near infinite diameter (where space-time is the circumference and is in motion where the diameter never reaches the circumference except in a Big Crunch)the seconds of arc can be subtended into planes which are in motion and must intersect because space-time is curved and not flat. And since each degree, minute or second of arc can be subdivided down an infinite number of times these planes which are non-parallel and in motion must intersect an infinite number of times making an infinite or approaching an infinite number of dimensions . These additional dimensions which exceed 26 would become more and more important as space-time constricts from near flatness to an infinite curvature point when a near infinite number of dimensions would occur which would be the n-dimensional state. The time component of four   dimensional space-time is what's curved by the effect of gravity caused by the inertial mass R ab. The effect of space-time curvature is in the numerator and the cause (inertial mass) is in the denominator revealing the curvature from the metric g a b based on the mass R a b and the curvature on flat Euclidian Space R a b c is R(t) or R d or time. When it is stated that space-time constricts near heavy masses such as black holes what is being constricted in space and time curvature going from flat (0)curvature to a point or infinite curvature is doing the constricting of space. Also space-time is a perfect fluid as shown in the diagram an area or region of Schwarzchild Space-time which is constricted s the event horizon of a black hole is approached has a moment which is equal but opposite where the magnitude is equal but the direction is in sum total opposite to the area(s) of constriction ,much as a circular rock is dropped into a pond where the rock is a perfect cone asymptotic to a spiral. The ripples   which go outward from the region of space-time constriction translate into the recently observed gravity waves or "g" waves observed near a black hole event horizon.

Time slows down as heavy masses are approached and speeds up as a vacuum is approached. Mathematically, R(t)=1/2 R g a b/R a b where R a b is a very large number R(t) approaches 0(zero) or essentially stops or goes very slowly in reference to the observer and R(t)=1/2 R g a b/R a b where R a b approaches zero(0)in approaching a vacuum making R(t)=R g a b/0 or approaching 0which makes R(t) approach infinity or move approaching infinitely fast in reference to the observer. This is why a traveler in space hardly ages with reference to the earth and comes back as approximately the same age when the earth has aged many years. For the traveler time appears normal but years evolve in what are perceived by the traveler as weeks or months. In the case of approaching zero gravity and approaching a vacuum state R(t)approaches 0/0 or everything including 0 but as 0 time doesn't exist 0 gravity either + or -1/2R g a b doesn't exist and 0 inertial mass is approached but doesn't exist. This is why traveling in a near vacuum state is like traveling quickly into the future and since the dark energy accelerating the push of galaxies away from each other is increasing arithmetically or geometrically time is speeding up.Time travel may not require the energy of a "Big Crunch"if a clockwise spiral that is increasing in diameter where the angular momentum from j to i where i=initial event and j is the final event that is decreasing is the mirror image of a decreasing diameter cone or spiral that is going counterclockwise where the angular momentum is increasing from the initial(i)event to the final event (j).With Schwarzchild Space-time the bi-conal configuration of space-time approaching the event horizon of a black hole would have a mirror component 180 degrees or pi radians from the decreasing spiral or cone on our side of the black hole. With positive gravity in the black hole space-time would have the infinitely dilated time component wrapping around space in one configuration such as counterclockwise while on the mirror side of space-time past the event horizon if space-time isn't just compressed toward an infinite curvature point has a clockwise rotation in a manifold where all events go in reverse or time's arrow is reversed. However as time=0 never actually occurs the point in the center of the biconcave configuration of space-time never actually hits zero just as the entropy(S) hits 0.29 and not 0.Despite this if true the time traveler would still be crushed at the event horizon unless the space he or she is traveling in along with the time traveler has compensated for all inhospitable environmental conditions including shrinking to near Planck Length without being killed instantly. Of course in a Big Crunch this could also happen to the time traveler.

As the energy density of matter rho incorporates inertial mass or the Ricci Tensor R ab with the approximation constant k~1/c^2 and since energy is a necessary component of time this Relativistic Effect incorporates 1/c^2 as R ab=$i\hbar\Lambda(\rho\ ab)$where $\hbar =$

$h\frac{'}{2\pi}representing\ Planck'sConstant\ or\ the\ wavelength\ of\ a\ photon(mass)(c)acco$

according to the deBroglie wavelength and the energy of a photon/(frequency) where high frequency reveals high energy and low frequency and long wavelength reveals very low energy.As a consequence inertial mass or R ab equals Planck's Constant(cosmologic constant which is the energy density of deep space with a "false vacuum" times i or the square root of -1 to incorporate the mathematic determination of anti-matter which is is strange matter with a hybrid mass delineated by mass/i as well as relating to the wave function of any point particle(r,t)in the Schrodinger Equation as

$$\text{ih}\Psi(r,t) =$$
$$-\frac{\hbar^2}{2mass}\nabla^2 \, times\, \Psi(r,t) +$$

$V(r,t)\psi(r,t)$ *potential energy and kinetic energy where* $\dfrac{\hbar^2}{2m\nabla^2 relates}$ *as the Hamiltonian Operator operating on Poten*

Hamiltonian Operator operating on potential energy and kinetic energy of a system
and ∇ *is nabla as compared to* Δ *which is delta. This relation with the Cosmologic Con* stant relates the energy density of matter to inertial mass as represented by R ab or the Ricci Tensor .The dimension of time is the fourth dimension of space-time .In the equation of everything space-time=space-time which is R a b c d=the infinite product for n=1 to infinity ½2^n(pi)R +R abc1/2 R g ab/rho(ab) or time R(t)=+ or -1/2 R g ab+^(cosmologic constant)/ Rab.d(n)=n dimensional state where n approaches infinity. The total equation of everything is actually $R(n)abcd = \Pi n = 1$ *to a where a* $\neq$

∞ *of* $2^n + \dfrac{1(\pi)(\omega\, j\to i)}{2^n(\pi)\omega'\,(i\to j)} + 2\pi[Rabc + or -\frac{1}{2}R\; g\; ab \div ih\rho\; ab\; where\; i =$

$\sqrt{-1}$ *.* $\omega = angular\; momentum\; for\; the\; spiral\; operator\; for\; the\; increasing$

Bi-conal surface or manifold and this is in the numerator of the spiral operator while ω' *is in the denominator of the decreasing conal surface of the spiral operator. Of course Of course*

That time will remain the same in all dimensions even if they are infinite only time's arrow can change or curve dilating or constricting angle between zero and two pi radians which can be subdivided down to infinity if space-time is a circumference of a circle(compact-i-fication of M Theory).Space-time is in motion and curved so there are an infinite number of intersections of each plane for each subdivision of each degree of arc so time is infinite as a dimension and space-time is infinite. Therefore nothing doesn't exist and time zero is an asymptotic value approached and never reached. Remember that the intersection of two planes defines a dimension. At the asymptotic value of "c" which is a boundary space-time constricts t *dual vector field*

$=seSE$ *for the covariant tensor R ab and the contravarient tensor* R^{ab} *which for time equals Rba or*

R^ b   a. The stress energy tensor T a b is stress between gravity's effect and the resistance of inertial mass which is represented in this case by the Ricci Tensor R a b while the space-time curvature metric + or –R g a b represents gravity and the energy equivalent of R a b as an inertial mass is *is* $4\pi R\; ab - 4\pi R\; ba\; or\; 4\pi(R\; ab - Rba) =$
$8\pi T\; ab\; or\; 2\rho\; ab. So\; incorporating\; the\; kc\; onstant\; approximatation\; k =$
$\dfrac{1}{c^2 makes}R\; ab\; into\; 4\pi\rho\; ab\; so\; R(t) =$
$+or -\frac{1}{2}\; R\; g\; \dfrac{ab}{4\pi\rho}ab\; or + or -$
$\frac{1}{2}R\; g\dfrac{ab}{}\; 4\pi\rho\; ab\; where\; the\; stress\; energy\; tensor\; is\; 2(4\pi\rho) and\; as$

$$gravity's\ effect\ is \frac{1}{2} R\ g$$

$$ab\ and\ -\frac{1}{2} R\ g\ ab\ it\ centers\ at\ R\ g\ \frac{ab}{\rho}\ a\ b.\ In\ a\ contraction\ of\ space-time\ R(t)$$

$$= R(t)$$
$$= R(-t)\ and\ the\ effect\ of\ gravity\ caused\ by\ inertial\ mass\ or\ it's\ equivalent\ with$$
$$the\ energy\ density$$

Energy density of matter as
ρ *the energy density of matter goes up with a constant total inertial mas*
mass and the vector fields

Effect go up with the difference being
4π *as a reciprocal in a contracting universe when we know the compactification*
of type IIA string theory or M theory

Type IIA string theory or M Theory is a circle of circumference
$n2\pi$ *radians so* 4π *radians is two revolutions* of space-time. Of course "The Equation
of Everything "becomes circumference(space-time)=spiral operator acting on
differential angular
momentum+2
π(*total Riemann Forces of Nature*). *As the eigenstates of the spiral operator* $\rightarrow$
infinity the spiral operator vanishes as it goes to zero(0) making
Circumference=2
πR *so the* Equation of Everything *mathematically becomes a circle. which is the
compactification of*
This circle is the compactification of string theory.This tensor version is actually
describing string theory in compactified form.Also one is to point out that the point of
maximum expansion the other to the point of maximum contraction involves "The Big
Crunch".The   point to maximum contraction has space-time constrict from an infinitely
large sphere of flat space-time to an infinite curvature point and
$2\pi\rho$ *is the energy density to point of maximum expansion with* $-$
$2\pi\rho$ *to point of maximum contraction. As the circumference decreases*
with a decreasing diameter indicating the sum of the r

Riemann   forces the decreasing diameter indicates that times arrow is reversed while
it increases as the diameter of Riemann forces increases. Note time's arrow only goes
backwards from positive time indicating
$n2\pi$ *radians to approaching but never reaching 0 radians as each second of arc*
can be subdivided down n infinite number of times

Down an infinite number of times   .Therefore time's arrow reverses in a Big
Crunch from

*2π radians to approaching but never reaching 0 radians in the compactification
of M Theory or type IIA string Theory*

Type IIA string Theory.

THE WAVE FUNCTION IN QUANTUM MECHANICS WITH REGARD TO TIME -_-
CHAPTER THREE

The Schrodinger Equation $-i\hbar\Psi(r,t) = -\frac{\hbar^2}{2m}\nabla^2\psi(r,t) + V(r,t)\psi(r,t)$ with $\hbar =$
$h\frac{}{2\pi}$ or h bar with $h =$
Planck's Constant reveals as nonstatic system of the wave function of
point particle r at time t with the Hamiltonian Operator

with The Hamiltonian Operator acting on the wave function of point particle r, t as the Potential Energy and V(r ,t)acting on the wave function of r ,t being kinetic energy of a dynamic system for the point particle. The Hamiltonian
is$-\frac{\hbar^2}{2m\nabla^2 acts}$ on thew ave function $\psi(r.t)$. and the static system where
standing waves develop as the the

a standing wave is time independent
.$E\psi = H\psi$ where E is the proportionality constant or energy of the state. $E\psi(r) =$
$\frac{\hbar^{2(-1)}}{2m\nabla^2} + V(r)\}\psi(r)$ where $\hbar = h/2\pi. i\hbar\psi(r,t) =$
$-\hbar^2/2m\nabla^2\psi(r,t) +$
$V(r,t)\psi(r,t)$ represents the potential and kinetic energy of the
wave function of the point particle at time tof a dynamic system

le of a dynamic system .The wave function of the probability distribution of the point particle ||(r ,t)||^2 in the n dimensional state is operated on by the Hamiltonian Operator from a=1 to n eigen-states of energy such that the n dimensional state of point particle r is
$dn\text{^}r\nabla^n$ and the Hamiltonian $H = -\frac{2\hbar^2}{2m\nabla^2 or \nabla^n in}$ the n dimensional state.
Again that's $-\frac{\hbar^2}{2m} \nabla^2$ operating on the wave function for weak perturbations
and energy levels with strong perturbations

Weak perturbations are dictated by the wave function in the Schrodinger Equation for the dynamic and static systems as a combined set of equations .In the case of a Big Crunch replace t with –t for the point particle and the energy will shift from increased kinetic energy to increased potential energy. .This can be done by the C.P.T. Theorem charge, parity time as in Quantum Field Theory indicating a symmetrical reversal of time and reversing time's arrow.

There is a C.P.T. (charge ,parity ,time)violation at the event horizon or every black hole as time dilates toward infinity or approaches stopping while space-time constricts toward zero as space constricts with counter-flow of gravity waves or g waves which occur as decreasing ripple in fluid space-time. The same or a similar occurrence could occur in "a Big Crunch" but in a black hole with Schwarzchild Space-time there should be a natural symmetry with space-time expanding toward flatness past the event horizon while it constricts toward zero as the event horizon is approached. As a result in black holes there may or may not be a C.P.T. violation .Note that one cannot measure the constriction of space-time effectively as "nothing cannot be measured" and no measurement is absolutely exact in quantum mechanics .In addition as the boundary of "c" is approached time dilates length shortens towards zero(0) and space-time constricts towards zero without reaching it.

Space-time constricts toward zero as the event horizon of any black hole is reached and curvature approaches infinity as the mass doing the curving increases. When did time begin? Time cannot be created or destroyed as space-time can neither be created nor destroyed as noted by the space-time continuum .So when the multiverse began or it's predecessors began what started it if energy cannot be created or destroyed and energy can be converted into mass by the Higgs' Boson or The Higgs Field which was been isolated and called the "God Particle" . Mathematically one can prove the existence of the Higg's Boson as a point particle r at time t where space-time is -1/2e-i n cot theta where i=the square root of -1 and theta is the trajectory of "The Big Bang" with the n dimensional state where n approaches an infinite number of dimensions proven in my previous book "Mega-physics II; An Explanation of Nature" .In the equation of everything space-time=space/mass or it's energy equivalent. As a result space-time=-

$$1/2e^{\wedge}\text{-}i \ n \ \cot \ theta = \psi(r,t) + H \ from \ 0 \ to \ n \ \ dimensions \ \|r,t\|^2 \ dn^r \ \nabla^n \ \ \div$$
$$\sqrt{\ } \ \hbar \frac{c}{8\pi G} \ \ where \ h\frac{c}{8\pi G} \ to \ the \ \tfrac{1}{2} power \ is \ Planck's Mass. \ The \ n \ dimensional \ state$$
$$of \ space \ requires$$

Point particle r at time t to exist. Without the point particle r at time t n dimensional space won't exist causing the impossible state of space-less ness .How does one show that r isn't the quantum bubble pre-Big Bang. The potential energy of the quantum bubble is finite but the mass of point particle r must be infinite because the mass in the denominator m=is infinite in the equation of everything rather than a subatomic particle. We know this because with infinite mass in the denominator space-time on the left side of the equation -1/2 e^-in cot theta becomes zero not approaching zero as in the event horizon of a black hole with a large non infinite mass in the denominator but space-time zero due to infinite mass of a point r at time 0.Infinite mass converts by Poisson's Equation converts to infinite energy density where the dual vector field is
$$\rho(\infty).\nabla(\infty,0) =$$
$$4\pi\rho(\infty). \ The \ only \ entity \ with \ \infty \ potential \ energy \ for \ which \ time \ and$$

are required to exist and the infinite potential energy comes from infinite mass for which The Higgs Field is considered as being self contained with massive Higgs Bosons. The wave function of point particle r at time t is also infinite without boundaries .Recall Schrodinger –

$i\hbar\Psi(r,t) =$

$-\frac{\hbar^2}{2m} \nabla^n \; \psi(r,t) +$

$V(r,t)\psi(r,t)$ *specifically in the n dimensional state where* ∇^2*is substituted with*∇^n *in the LaPlacean Oparator*

$$\nabla^n.$$

An operator where n approaches an infinite number of dimensions which eventually subdivides down to where dimensions are so immeasurably small that they appear to mimic the zero dimensional state which does exist(infinite number of parallel planes) . Still this isn't nothing as nothing doesn't exist and space-time is a continuum .If the planes are all parallel to each other with no curvature you end up with the zero dimensional state which is a single plane or infinite number of planes(totally flat space-time). A single manifold or an infinite number of manifolds in the zero dimensional state may actually be the quantum ground state reflecting the Higgs Field(with tachyons as a subcomponent)with almost infinite potential energy Now we know nothing plus everything is everything. The infinite potential energy and mass of the Higgs Field was always there as it was essentially creating an infinite space-time continuum and existing with innate symmetry as a continuum of inter-fitting spirals or cones in a solid plate of infinite dimensions while acting as a wave rather than a particle so the plate was composed of waves inter-fitting spiral geometry.A mathematical proof that an infinite number of cuts of the circle which is the compactification of IIA String Theory can show that if one visualizes a circle with two cuts R1and R2 where their intersection represents a dimension and the circumference is of course space-time then represent space-time curvature or dr^2 asθ *if*$\theta = 0$ *degrees as in totally flat space* $-$ *time or the zero dimensional state as* $dr^2 \rightarrow 0 then\theta \rightarrow 0$ *and in this case* $dr^2 = 0$ *radians. In this case* $R2 =$ $R1$ *so this would mathematically show the zero dimensional state as if indeed* the Higgs Field can exist without space-time then this would indicate a possible state of the Higgs Field. Of course as space-time is a continuum(without beginning or ending)as Einstein stated, this state is a super-symmetric lattice structure totally without any perturbations weak or strong and this appears to be impossible(beyond the comprehension of mankind),but it is mathematically proven to exist so it is not nothing by definition. To show non-flat space-time would be indicative that space-time is in motion(as it is)rocketing outward at 2.2x10^35times the speed of light "c"meters/sec with a jet steam current of space-time traveling at (pi)times the speed of light pulling which along the push of Dark Energy causing the expansion to accelerate geometrically.

Space-time acts like a "perfect fluid "with currents ,eddys and backwash so it can OSCULATE(touch)other space-time an infinte number of times with unit tangent vectors(which increase as the subdivisions of the angles of the osculating planes get closer and closer to 0 degrees or 0 radians which smaller and smaller);while the Orbifold in String Theory is still within the Hilbert Space limit. Indeed space-time develops a granular or grain property below this which is far far under Planck Length(10^{-33} cm)and composes what this author calls the Super-Brane compactifiable up toward infinity as it becomes closer and closer to being infinitely small. Of course space-time curvature in this case is skewed toward asymptotic flatness as the zero radian or degree limit is approached as an asymptotic limit, and this can be showed as the limit as space-time curvature approaches zero of ds^2 which is derived from the line element ds^2=dx^2+dy^2+dz^2-c^2dt^2+dr^2 where dr^2 approaches zero and dr^2 is space-time curvature. This limit must be

multiplied *by 2π to describe the circle of String Theory*

so we arrive at $2\pi \int_{\infty}^{0} ds^2$ *whichis* which is

$2(\pi s^3 /$

$3|\infty$ *to 0 or ∞ indicating an infinite number of unit tangent vectors or dimensiondd* dimensions mathematically proving that there are in infinite number of dimensions rather than the 26 non-compactified described by String Theory. There are an infinite number of planes from the subdivisions of each degree of a circle(compactification of String Theory).If these planes represent space-time which is curved and in motion they are never parallel and will form a dimension with each intersection but if these planes are all parallel they will never intersect representing the D-0-Branes as they represent the zero dimensional state of an infinite number of planes. These parallel planes are D-0-BRANES and there were an infinite number of them. There was a weak perturbation in the D-0-BRANES causing a CENTRIFUGE EFFECT CAUSING ALMOST ALL THE D-0-BRANES TO COMPACTIFY TO SUB-HILBERT SPACE SIZE AND BELOW PLANCK LENGTH TO FORM THE SUPER-BRANE. Of course, zero dimensional space is still space ;it isn't space-less-ness and an infinite number of D-0-Branes or parallel planes are still space. Time is the part of the space-time continuum that changes space constricting it and dilating it. Time always existed and will always exist as the first event in the formation of the multi-verse could have happened without time; consequently is there a first event? Think of time as a container or a superset in which events are members of the set whether times arrow goes forward and backward. In this sense the set should be Abelian(related members of the set but not in that the ordering of the sequencing of events is important.)The container is always there regardless of whether or not it has contents as in the" space-time continuum" however bosons and tachyons (which display + and – mass and therefore the effect of gravity and possibly anti-gravity) should theoretically have a beginning while the infinite parallel planes of zero dimensional space or D-0-Branes wouldn't have a beginning as they are a container which describes an "infinite space vacuum state "as space-less-ness is impossible and the "Law of Conservation of Dimensions "states that all the dimensions in a system must be a constant.Indeed,tachyons may not have had a specific beginning traveling above 3x10^8 meters/sec until a sentinel tachyon broached the speed of light barrier to transform the fermions in the fermionic vacuum state to the first boson while the time paradox of times arrow oscillating between forward and backwards started the harmonic that ransformed the infinite parallel planes into a vortex of space-time where the center of the vortex is the string dimensions and the outer rims are the macroscopic dimensions.This indeed explains Einstein's idea broached in 1912 that space-time has a spiral configuration as we are not limiting space-time to our asymptotoically flat universe but to the multi-verse in the space-time continuum(no beginning or end) which is why the spiral fractal equation solves the configuration of space-time for the quantum ground state l n 0 and l n 1 =0 in a relativistic universe as pointed out in this author's first book "Megaphysics ;A New Look at the Universe"(2003) where the integral of du/u explains the Hubble Expansion Coefficient

$\int \frac{da}{a}$ *as well as the solution being* $\ln a$ *which from* $-$

∞ *to* ∞ *reveals* 0 *or the ground state not just of this universe but all of the multiverse* the multiverse as the centrifuge effect forms a vortex of space-time which is spiral.The time element of tachyons in the infinite space vacuum state of D-0-branes reveals – infinity—∞ *as times arrow is reversed here but instead of time*

　　　　constricting space toward a point of infinite curvature
with infinitely dilated time or tiime
$= \infty$ *negative time dilates space toward infinty making space*
$-$ *time totally flat in the vacuum or fermionic state with infinite parallel planes.*
Relativity illustrates a conversion of dimensions as a mass approaches the speed of light such as the progressive shortening of length as per the Lorenz Transformations however this doesn't mean that width, height and time are destroyed. Time dilates but it towardsardstof other dimensions ;they can dilate or constrict along with time. Mass requires space;but proving that space can be created or destroyed seems logically wrong as mass cannot empirically exist without space. This doesn't mean that space cannot exist without mass as in the infinite planes which existed before "the first event". An event results from a standing wave　which resulted from the successive dilation and constriction of time acting as a Harmonic Oscillator. This Harmonic Oscillator caused the spin which caused the centrifuge effect and formed the spin-2-vector bosons which acted upon time to curve space(gravity).This is why the HIGGS-Boson-Tachyonic Field(s)form a vortex as per the tensor virial theorem(see second Ontologic Proof included from Journal of Mathematics later in appendix).The weak perturbation came from DIFFERENTIAL TIME AS TIME'S-ARROW IS OPPOSITE BETWEEN TACHYONS AND BOSONS AND TIME ALWAYS EXISTED AS EVENTS ALWAYS EXISTED. The reversal of time's arrow with forward time's arrow causing time to dilate and constrict which caused a HARMONIC PARADOX in the dimension of time which wrapped around space causing it's curvature as a repeating HARMONIC OSCILLATOR WHICH CAUSED THE PERTURBATION FORCING THE HIGGS FIELD INTO A VORTEX　CAUSING THE 26 COMPACTIFIED DIMENSIONS OF STRING THEORY REDUCED TO 10 OR 11 TO HAVE HAD LESS OF A SPIN(than the other super-compactified infinite number of dimensions). THIS IS WHY THE SPIN 2 VECTOR BOSON DESCRIBES THE EFFECT OF GRAVITY.OF COURSE THE REMAINDER OF THE INFINITE DIMENSIONS GRAVITATED DOWN AND DOWN AND DOWN TO INFINITELY SMALL REGIONS ALL INCORPORATED IN THE SUPER-BRANE FROM THE CENTRIFUGE EFFECT RESULTING FROM THE SPIN. With 524,288 equations in nature which are the permutations of "The Equation of

Everything" there is still much work to be done by the Scientific Community before an event happens. Regardless of this ,an explanation of Einstein's Equation of Relativistic Gravity as one of Einstein's Field Equations explains how the Einstein Tensor G ab=0.The field equation is G ab=R ab + or -1/2 R g ab=8(pi)G/c^4 T ab or ~8(pi)T ab. Space-time curves inward with positive gravity and curves outward with anti-gravity as in Dark Energy as described by the Cosmologic Constant. R ab describes the inertial mass or Ricci Tensor and T ab is the Stress Energy tensor between gravity(anti-gravity)and inertia. In the case of a pure 50:50 mix of gravity and anti-gravity space-times curves outward and inward cancel making space-time perfectly flat as in the infinite number of parallel planes of the infinite space vacuum space of the zero dimensional state before strings were formed by the oscillating time paradox. Of course in this case there would be a massless state as space-time (flat) is totally massless. As a consequence to this +1/2R g ab and -1/2R g ab cancel out to 0 as without mass there is 0 gravity and 0 anti-gravity making the Einstein Tensor=0 as the Ricci Tensor =0 , gravity and anti-gravity=0 and the Stress Energy tensor must also be 0.The boson/tachyon mix causing the time paradox had 0 mass also as the Bosons and Tachyons had equal but opposite mass canceling them out until a weak perturbation tipped the mass into existence as the motion of tachyons upset the boson balance. Tachyons may have always been in existence forming bosons when the speed of light(c)was penetrated to the downside and this may have been the perturbation that caused the oscillation in time which formed the existence of strings. Tachyons are an important component of "the God Particle "and may have pre-existed bosons by an infintescemally small period of time.

CHAPTER FOUR :"IS TIME TRAVEL POSSIBLE?"

In an expanding universe with Dark Energy acting as an anti-gravitational effect pushes galaxies apart from each other faster and faster .Regardless of the velocity space-time will always outrun the Dark Energy in an expanding universe unless a singularity is reached when the expansion exceeds the velocity of space-time greater than πc. *This singularity will rip space —*
time in the fabric of the confluence of space —
times in the multiverse forcing a back flow of the expanding space —
time to a constriction of space —
tiem along with all the 750 billion galaxies. This is the Big Crunch*effect*
*   where gravity which travel over c because it is the curvature*

Curvature of space-time increasing at over c rather than the flattening effect of antigravity which also travels over c. So in an open flat expanding Friedmann type II universe any time traveler will travel into the future faster than the external environment around him.The faster he travels in relation to the external environment the faster he travels into the future until "c"is approached when space-time constricts rather than flattening and here time dilates or approaches being infinite curvature as in a point ;here time almost stops or goes infinitely slow for the time traveler and rushes by for the external environment. This is analogous to the tip of the cone in Schwarzchild Space-time where it goes from flatness(asymptotic)to ward an infinite curvature point. So to the time traveler he or she can go years into the future of the external environment with respect to himself or herself. This is where antigravity predominates and space-time curves outward until the velocity approaches "c" the speed of light boundary when space-time constricts toward an infinite curvature point without reaching it. So the next question is obvious   ;is time travel backwards possible ?The answer is yes. The effect would be like watching a movie in reverse faster and faster relative to the time traveler as one goes further into the past. This can be accomplished at great cost   ;a pyric victory as the energy of a Big Crunch would be required in a massive implosion of all the galaxies back toward a not the quantum bubble. This would make time travel meaningless as the time traveler would eventually be crushed in the singularity is time travel in reverse is too fast. There would have to be a force equal but opposite in the time vehicle to the force of the Big Crunch or 6.75×10^{34} erg to prevent the time vehicle from being crushed. In this case thermodynamics would show entropy decreasing instead of increasing down towards 0.29 which is the entropy of a black hole according to Steven Hawking and as billions of degrees would have occurred after the Big Bang as a heating process this would be reversed down to 2.74 degrees kelvin in the time traveler's vehicle so all matter would become Boso Einsteinian Condensate until the new quantum bubble is approached where 252 separate states of matter would occur including solid radiation , liquifying radiation due to the extreme pressure and the extreme curvature inward of space-time as in a black hole with an extremely heavy mass(gravity).Heavy mass, heavier gravity(more inward curvature od space-time)would cause time to dilate toward infinity as

infinity$\infty \div large\ finite\ number \rightarrow$

$\infty\ with\ space-time\ approaching\ infinite\ curvature\ for\ the\ gravitational$
$component\ in\ the\ numerator\ but\ notreaching$

Reaching it and the denominator is a large non-infinite number so time dilates toward infinity without reaching it .R (t)=constricting space with extreme curvature   as in a black hole/extremely heavy non-infinite mass. There is a threshold mass which would cause infinite gravity or space-time constricting to a point that almost infinitely small .This may be hard to visualize but could describe the Higgs Boson prior to the formation of all subsystems creating the multiverse in the most stable confirmation for the multiverse; a Calabi - Yau manifold(surface)which is six dimensional. This would truly be

time$\rightarrow$

$0\ or\ the\ beginning\ of\ a\ cycle\ of\ time\ which\ had\ time's\ arrow\ go\ foward\ slower$
$and\ slower\ as\ the\ implos$

Implosion progresses until it dilates toward zero when mass in the denominator approaches infinity. But when does time's arrow go backwards? There must be a huge conversion of the potential energy of that infinite mass into kinetic energy. Unfortunately, this kinetic energy must be negative not positive and this mass must be negative as in a black hole beyond the event horizon .As dark energy is anti gravitational and comes from antimatter the mass of the antiparticles must not be normal positive mass despite the fact that in the Hadron Collider at Cern anti particles have a primarily positive mass; but these are isolated particles as antimatter lack cohesion so the mass must be a hybrid between + and – or mass/i   where i=the square root of -1.Negative energy seems to be past our technology and negative mass doesn't appear to be harness- able as antimatter lacks cohesion however in theory R(t) becomes R(-t) when the derivative operator of a dual vector field from Poisson's Equation becomes

$\nabla \left(\dfrac{mass}{i}, space - \right.$

$time\ curvature\ metric\ or\ gravity\ where\ the\ curvature\ uncoils\ toward -\infty \left. \right) =$
$-4\pi\rho\ where -\rho\ reflects\ negative\ energy\ density\ of\ the\ hybrid\ mass;\ \dfrac{mass}{i}.$

Going backwards in time is possible but would most likely destroy the time traveler and possibly revert everything else to it's initial state with only the Higgs Field or Boson.It should also be noted that at some future date perhaps in 500,000 years technology may be sufficiently advanced to turn time's arrow backwards without killing the time traveler.If our civilization gets advanced enough for interstellar travel and encounters a "worm hole"(first demonstrated by Albert Einstein)and if this worm hole is in the middle of a Kerr Loop made of superstring material and the time traveler enters at the terminus of the Kerr Loop initiates going in a Region of Space-time where time's arrow is reversed then enters a worm hole, that traveler would leave the worm hole parsecs away in space and sometime in the past. Unfortunately, worm holes are only theoretical and there has been no empirical evidence of their existence. Also if a boson/tachyon interaction could occur curved space could become a space-time vortex where the tachyon component of the vortex could propel a traveler into the past.(tachyons are theoretical particles which go back in time but have only be mathematically predicted and never isolated)This may or may not be harmful for the time traveler. Finally a word of warning for those in the future who may develop this technology ;if time's arrow is reversed and a time traveler gets into some distant or recent past; the time traveler must stay in the past and not attempt to return to his or her time of origin as this will disrupt space-time possibly causing a rip ,tear or tidal wave effect with a flush effect and a "Big Crunch ",so a universe can still be destroyed by time travel into the past. If individuals return to the present with the time traveler the disruption of space-time would be more catastrophic ;so it is prudent to avoid time travel to the past until we have mastered singularities.

Space-time is virtually never totally flat as an absolute vacuum without energy or matter or mass-less-ness is impossible as one assumes that photons which comprise electromagnetic radiation have mass albeit miniscule($3x10^-18$ eV/c^2 or smaller)yet enough to be attracted to black holes and follow space-time curvature metrics being bent by gravity .In actually space-time is curved by mass; mass does the curving and the dimension of time is what's actually been curved. Visualize a ball with an infinite diameter being Euclidian space for the first three dimensions and time is a cord which around black hole event horizons approaches being infinitely long or dilated (time almost stopped or time infinity).The infinitely long cord wrapped around the sphere of infinite diameter will squeeze the ball down to a point which is virtually immeasurable while space-time has infinite curvature and this is the infinite length of time constricting the space of the ball down to almost nothing which is why at black hole event horizons and just under "c" the speed of light space-time constricts toward zero while time is dilated to near infinity by the Lorenzian Transformation time=time(initial)/1-'v^2/c^2)^1/2 which us initial time/0 at v=c or infinity or total dilation where time stops .In a vacuum where the BMR makes 2.74 degrees kelvin in deep space space-time is virtually flat without curvature and it's this predominance of virtually massless space between galaxies that makes the total space-time curvature of this universe at 10^54kg a value of $5x10^-29$ radians where 2 pi radians is 360 degrees. The curvature would never be absolutely zero because absolute zero is a singularity point which is never reached . Flat time is a uniformly peppered continuum along space which is a compilation of cones and inverted cones acting as a solid matrix .Einstein stated that mass creates space. This statement is incomplete and only partially correct depending on the reader's point of view . Energy or the energy equivalent of mass create space and this is only because energy as photons have a miniscule mass and the photons create space while the mass of the photons curve space such a tiny amount it's virtually immeasurable   .Remember that nothing cannot be measured, nothing doesn't exist and there are no exact measurements only ranges according to the

Heisenberg Uncertainty Principle .Flat Minkowski Space-time is the peppering of flat Euclidian space being peppered by the dimension of time and as energy isn't massless Minkowski Space-time isn't totally flat but may have curvature of 10^{-250} radians or less. Space-time=space/mass. If mass is zero(0) space-time=space/0=∞ *and space − time is infinite as a continuum without beginning or end. In reality masslessness is impossible so*

As mass-less-ness means photons have zero mass and there would be no energy or mass which is nothing which is impossible(beyond the comprehension of man)so we assign a value of epsilon(a very small value)which is added to 0 mass which makes space-time a huge number just under infinity and space a huge number just under infinity. Actually the inertial mass becomes the energy density of matter in the denominator as e=mc^2 is only an approximation and is even less accurate in the calculation of the energy of photons. So again space-time=space/energy density of matterρ *from Poisson sEquation.*

When a traveler goes into outer space his internal clock slows down relative to the external environment is he or she was traveling near the speed of light. Where time is dilated so returning to earth will show the environment in the far distant future. The same or similar thing will happen near the event horizon of any black hole. However to travel backwards in time you need to infuse negative energy from hybrid mass from antimatter into an existing implosion from a Big Crunch. This appears as mentioned previously to be very difficult with our technology. So if the Big Bang was preceded by a Big Crunch of another universe rather than the collision of touching of two membranes ,in order from time's arrow to reverse you would need a universe whose time sequencing is the OPPOSITE to ours so time goes from very negative to approaching but not reaching time zero at 10^{-43} sec prior to "The Big Bang" .In other words in that universe probably composed primarily of anti-matter with variable or hybrid mass or mass/i would go from a Big Bang whose conformal time is just the opposite in configuration or spin to ours having the point of maximum expansion of our universe equal the point of maximum contraction or "The Big Bang" of the previous universe. As space-time curvature is the opposite with anti-matter to ordinary matter one may conclude that it was a universe where anti-matter had positive gravity instead of anti-gravity and regular matter had anti-gravity .In a sense it was a MIRROR or CHIRAL UNIVERSE WHERE THE CONFIGURATION OF SPACE-TIME WAS A MIRROR IMAGE OF THE MANIFOLD IN OUR UNIVERSE.

CHAPTER FIVE
__CONFORMAL TIME

With the cosmic Inflation theory by Dr .Alan Guth shows the relationship of every point in space-time remains the same from the point of inflation to the point of maximum expansion like an expanding balloon. Conformal time relates to conformal space as the fourth dimension but as the relationship between every point in time and every other point of time remains the same or is inflating at the exact same rate (although Dark Energy is accelerating the expansion and may cause the Popped Balloon "scenario this author mentioned in my previous book "Mega-physics II; An Explanation of Nature". The popped balloon scenario would be an abrupt start to the rupture of the space-time of our universe in conformal time and would most certainly trigger a Big Crunch which would be very very fast possibly Planck Time 10^{-43}sec.Conformal gravity says that for each and every target mass doing the curving on Conformal Time in Conformal Space the superstring of Conformal Time curves miniscule to almost an infinite amount by masses varying from the photon where it approaches 0 mass to the largest Black Hole(probably the yet undiscovered Black Hole which was the site of "The Big Bang "and might actually be the center of mass for this universe which would be the center of rotation which is decreasing and has been decreasing since "The Big Bang" as the accelerated expansion has been increasing and accelerating either geometrically or arithmetically. Kurt Godel originally came up with "Godel's Rotating Universe "which would logically rotate around the site of "The Big Bang "which is again the center of mass .The universe has a mass of 10^{54} kg and every mass must have a center of mass according to Sir Issac Newton so there must be a center of mass for this universe and since mass curves the time component of space-time there must be a center of gravity and it must be the event horizon of a huge supermassive black hole where "The Big Bang "occurred. Conformal gravity being the curving of conformal space-time caused by mass that has the same fixed relationship or ratio of distances with every other mass would make it difficult to show a center of mass for the universe. The near iso-tropism of this universe showing observational symmetry from any vantage point makes it even more difficult to find a center of mass for this universe ,but if there was no center of mass and no supermassive undiscovered black hole where "The Big Bang "occurred then how did the Big Bang occurred without causing a Black Hole? This universe does rotate   ,it must rotate and it does so less as less in space-time for this universe expanding into the space-time of other universes. If the rate of cosmic inflation is increasing geometrically from dark energy the balloon could "pop" sometime in the future. If there is "a Big Crunch "and if "The Big Bang "is correct and not "The Inflation Theory" it will take longer than Planck Time to implode this universe and there will be warnings. First the red shifts in the Doppler will start to shift from infrared toward ultraviolet and secondly the almost immeasurable rotational component from "The Big Bang "that uncoiled with the accelerated expansion of space-time (of this universe) into space-time(of other universes)in the perfect fluid of space-time will start to increase again ,first for some time the rotational component around the black hole at the spot of "The Big Bang" will still be in perceptible and only a decrease in the red shift will be noted but later the rotational component of the expansion-rotation of this universe will be measurable and will increase first slowly then faster and faster until the spiral component of space-time will become predominant over that of almost complete asymptotic flatness .In this case "The Big Crunch" will be a flush down to another quantum bubble. The only thing that doesn't have a center of mass or gravity is space-less-ness or nothing. Nothing doesn't exist and space-less-ness is the only true boundary as "c" the speed of light boundary is actually a "false boundary"-as space-time travels over c and perhaps tachyons. Recall that the speed of gravity is the speed of space-time in actuality and that it is mathematically postulated to be

πc to the point of maximum expansion and πc to the point of maximum contraction

Contraction .This again is due to the compactification of type IIA string theory being a circle and also M Theory which is all five string theories .The circle has space-time as circumference and the Riemann Forces as the diameter. The diameter never reaches the circumference except in a Big Crunch singularity and these forces travelled at c just after the Big Bang until stopped by almost infinite inertial mass according to the Lorenz Transformations .As a result Riemann Forces didn't hit "c "but space-time exceeded "c" and out ran it by π according to the $\int_0^\infty dv/dt$ where dv/dt is the velocity of space-time so

dv/dt=$c\int_0^\infty dv \int_0^{2\pi} d\theta$ which is $2\pi c$ as the cycle of space $-$ time in the circumference of the circle which is the compactification of M Theory.

Here the limits of integration for d V are time=0 to time=infinity so infinitely large space going from t=0(flat space-time)to t=infinity(infinite curvature or a point)forces c(v from time 0 to time infinity)

θ or over all time $c\theta|$ 0 to 2π radians. So the velocity of the current of space $-$ time is $+$ or $- \pi c$ which over infinte time accelerates toward infinity.

$$\frac{dv}{dt} =$$
$$c \int_0^\infty dV \quad \int_0^{2\pi} f(\theta)d\theta . \theta \text{ and in the cycle to maximum expansion and contraction}$$
$$\text{the first of the double integral convert}$$

converts from infinity to 2
π radians or the circumference of the circle which is compactified type IIa
string theory as

2π radians is the maximum circumference of the circle. As long as the Riemann
forces in the

diameter are finite the circumference is only two pi radians or 360 degrees and the infinity of space-time velocity is bounded as under infinity and determined to be $2\pi c$ or πc to maximum expansion and πc to point of maximum contraction with reference

to space-time from other universes. Remember we're dealing with 2 pi radians as the circumference and 2 pi "c" as the velocity of space-time to point of maximum expansion then back to maximum contraction being pi"c" each way. Basically

$dx/dt = c \int_0^{2\pi} d\theta = c$

$\theta | 0$ to 2π radians where the circumference'sangle are spacetime curvature

so $c\theta$ is $c(0) - c(2\pi)$

$= -2\pi c$ for the total cycle to maximum expansion and maximum contraction.

$Dv/dt =$
$\iint_0^\infty dV \, d\theta$ where the limits of the velocity of space $-$ time are 0 to infinity and for $d\theta$
(angle of spacetime curvature along cicumference of

Circle with Riemann forces as diameter that are finite; $V = 0 - (-\infty)d\theta = \infty \, c \, \theta =$
$-2\pi c(\infty)$ which because of the double integral indicates an
expansive current of πc in an infinite

Ocean of perfect fluid space-time .So as a result when space-time is expanding into total space-time the expansive space-time velocity is πc and the constrictive space $-$ time velocity is $-\pi c$ and $\pi c - \pi c$ which is a net velocity for this universe $\rightarrow$ 0 if a Big Crunch occurs. This is the velocity of the currant of space $-$ time in infinite space $-$ time which has infinite total velocity which is why space $-$ time within the confines of
one of many universes approaches 0 if cyclical as a currant in total fluid

fluid space-time .Note now there is a valid explanation as to why the expansion of this universe is accelerating either arithmetically or geometrically. Dark Energy is anti-gravitational and appears to have far more force than measurements indicate. This is because space-time from our universe or manifold is traveling at
πc and accelerating but it is accelerating as a current into space $-$
time traveling at near infinite velocity from the multiverse
and this effect is sucking or pulling

pulling space-time as a current(I) into space-time which is traveling faster causing the suction effect of a small compact area of space-time being pulled into a huge ocean of space-time where the space-time that is being pulled or sucked is from our manifold(universe)into the space-time of all other manifolds in the multiverse traveling at infinitely larger velocities .This pushing from Dark Energy and the pull from Space-time from other universes or manifolds and all outward in the massive accelerating expansion. This effect has been since "The Big Bang "and explains why the mutual repulsive force of anti-particles or anti-matter in the quantum bubble was so huge with Dark Energy, The pull of space-time from other manifolds on our manifold or universe acting on the push of Dark Energy's effect of anti-gravity so the same stuff was being pushed outward and pulled outward simultaneously.

SPECULATION CHAPTER SIX

Why would the velocity of total space-time approach infinity even with space-time being infinitely large? Velocity means traveling from point A to point B and while currents in space-time can travel at velocities A to B (as in out universe) what is point B for total space-time to be traveling at near infinite velocity? Answer is just an educated guess. Recall this author stated that the only boundary is space-less-ness and that space-less-ness is impossible (that which is beyond the comprehension of man).If this boundary was breeched a near infinite distance away either in another manifold or universe a distance away so large it's almost immeasurable(decillions of manifolds or a googolplex of universes away)space-less-ness would pull space-time in toward it at a near infinite velocity and this would be point B .Point A would be the initial encounter for all of the multiverse or any supers-system which subordinates the multiverse. This was mentioned as totally flat space with time as a confluence of microdots in a lattice of inverted cones and cones inter-fitting into a matrix .If this did happen the tear of space-less-ness would increase as it fills with space-time until it either reaches a steady state or collapses to a point or microdot of submicroscopic dimensions almost infinitely small with all the approaching infinite dimensions being compressed as the one second of arc of the circle that is the compactfication of type IIA string theory or possibly M theory .This steady state or equilibrium would be determinable by a decrease in velocity of space-time from infinity to some lower amount and this would slow the velocity of the current of space-time in our universe ever so slowly which would slow the left shift down slightly but probably too small to measure accurately. It would also dilate or slow time slightly but certainly not large enough to be detected.

THE THEORY OF THE MOBIUS CHAPTER SIX a

If time were a loop or Mobius Strip then point A(origin) and point B(destination will be in close proximity forming a circle that is infinitely long with a circumference which represents

$n2\pi$ *radians where n* $\rightarrow$

∞. *The diameter can fluctuate from near* ∞ *to near* 0. *If the diameter* $\rightarrow$
0 *the circle becomes a point with infinite space* $-$
time curvature but the action (s) *of the mobius would be that of a repeating*
                    or oscillating

Circle   which can be deformed by internal currents of space-time like the accelerated expansion of space-time in our universe which has a drag emanating from trillions of black holes. This scenario is possible with Kerr loops of superstring material and if the oscillations approach an infinite number a finite number of loops can appear in the circumference emanating from mass creating a drag on space-time curvature so the oscillation can be dragged as they would be from the curvature inward from a black hole event horizon. The advantage of this idea is that an infinite velocity of space-time would be from A to A which would explain a Big Bang-Big Crunch-Big Bang scenario etc. with the drag exceeding the pull of the dark energy to induce space-time constriction rather than expansion which would make the velocity of space-time=n2

πc where 2 describes the cycle to maximum expansion and maximal constriction or contraction

Where
n→
∞ velocity without having a space − less −
ness singularity a googlplex or more universes away but then you don'thave an elegant explanation of the acceleratin

Explanation but you then don't have a clear explanation of the accelerating expansion of space-time and the galaxies in excess of the anti-gravitational effect of dark energy. It may cause the possibility of time travel backwards in a Big Crunch from another universe however, so the theory of the Mobius cannot be dismissed but may be considered incomplete although again the compactification of type IIA string Theory is a circle and possibly M Theory. So the velocity would be from A to A a near infinite number of times.

Please note that a rip or tear in space-time or an miniscule area of space-less-ness a huge distance away (trillions of universes or greater) may not directly affect our universe as was previously mentioned a steady-state could occur which might just cause the acceleration of space-time in our universe to slow slightly when the Big Crunch occurs in another universe because the pull of space-time from other universes on the space-time from our universe would slightly decrease and therefore the acceleration of space-time in our universe to exibit a slight drag in the pull. If there was a Big Crunch before our "Big Bang" the space-time that crunched from that universe would be the SAME space-time that's exhibited in this universe not space-time from other universes. So empirical measurement of the velocity of space-time is important to determine areas of drag vs. pull and their indirect effects.

CHAPTER SEVEN

Vacuum energy

IT HAS BEEN POSTULATED THAT VACUUM ENERGY can cause the accelerated expansion of the 750 billion galaxies away from each other but the amount of energy incurred from a relative vacuum; containing space and almost nothing else is not nearly sufficient for the accelerated expansion of galaxies apart from each other even while being pushed by Dark Energy. An implosion from vacuum energy might implode a galaxy or possibly even a universe into a "Big Crunch" but would not accelerate space-time to a near infinite velocity from πc towards ∞. *However an area of space $-$ less $-$ ness in a distant universe would accelerate space $-$ time in our universe to go from $\pi c \rightarrow$ ∞ until a steady state is reached and the velocity has slowed by the sealing of the space $-$ less area with space $-$ time and would cause the current of space $-$ time in our universe to accelerate either arithmetically or geometrically the laddder*

latter would cause a "Big Crunch "in our universe which if inflation was true could cause more of a Big Pop collapsed matter and energy to another quantum bubble in 10^{-43} seconds.

The is some empirical evidence using CMB data and what is termed Dark Flow by the Planck Team at U.S.C by Dr.Elena Pierpaoli[1] and a pull on the B.M.R.as illustrated by Laura Mersini-Houghton physicist at University of South Carolina at Chapel Hilland Richard Holman professor at Carnegie-Mellon University predicted anomalies in the BMR existed and were caused by the pull of other universes[2] in 2005.The findings of both these physicists postulate a near infinite number of universes so a "Big Crunch" singularity from an area of space-less-ness(quantum mechanics says it must happen, is happening and will happen") and because of this a universe almost a googolplex away can accelerate space-time towards infinity from *πc causing a huge pull in addition to the push of Dark Energy which was*

exaggered (scientific measurement show Dark Energy considerably less than it really is)is due to the extreme pull of space-time although anti-particle anti -particle repulsion in a spot $1)^{-24}$ cm in the quantum bubble causes the accelerated expansion.

CHAPTER EIGHT
BOUNDARIES FOR CONFORMAL TIME OR CONFORMAL GRAVITY OR WEYL'S CONFORMAL TENSOR

Weyl's Conformal Tensor C ab^ cd or C ab c d or –C dc ba etc registers conformal gravity or conformal weight in reference to space-time or it's curvature metric. Cab^

cd$\rightarrow R\ ab^{cd} + or - \varpi(t)e + \frac{or-1}{2\phi w}$ which in a bosonic field yields e-

^(+1/2ϕ) which yields true fermionic weights of $-$

$\frac{5}{8}$ and $\frac{3}{8}$ with the true fermionic vertex weight (ferminics $\rightarrow$

0)in the vacuum state or if an area of space $-$ less $-$

ness appears in a rip or tear in space $-$

time as in a Big Crunch in another universe. $V - \frac{1}{2} = \mu^\alpha\ e^{-\left(\frac{1}{2\phi}\right)\int \alpha}\ e - ik.x$ or

$\mu^\alpha\ e^{-\left(\frac{1}{2}\right)}\ \phi\ \int \alpha\ e^+ or - \frac{1}{2\phi}$ so the conformal weight is $\frac{5}{8(-1)} + \frac{3}{8} + \alpha\ k^2$ or $\frac{5}{8} + \frac{3}{8} =$

1(αk^2 and k^2 represents space $-$ less $-$

ness with the Vertex function having a conformal weight = 1 =

$\frac{3}{8} + \frac{5}{8}$.γ^μ $k\mu c\alpha =$

0 where c α is the conformal weight of the ferminonic field and $\gamma^{\mu k\mu}$ is the true fer

Fermionic vertex weight in a bosonized field 3 such as the Higgs Field. In a singularity such as a Big Crunch or rip or pop in space-time V V=[Q

BRST,ϕ] =

the null state of space $-$ less $-$

ness WHICH FAILS TO COUPLE WITH ANY REAL STATES. What is termed the Bose Sea Bose Sea may also be a coupling of the Higgs Field(god particle)with the perfect fluid of space-time. Note Bose Sea is analogous to a sea of bosons   which when related to fermions also relate to gravity or the space-time curvature metric caused by mass. Mathematical ghosts have to be inserted as "fudge="factors to make the vacuum or null state exist such as fudge a ghost sector with a conformal weight of 3/8 and in insert as a fudge factor to make a matrix element into a sea of bosons go to 0 a fermionic matrix must be inserted such that <...V-1/2....V-1/2....>=0 rather than approaching zero but never hitting it,A mathematic fudge factor must cancel the -1/2 charge with a charge of +1/2 in the ferminoic matrix of the vacuum state and V1/2 and V-1/2 must be anti-communtative. The charge related vertex field Q   BRST|R>=0 and the BRST charge in this vertex field must vanish in zero matrix or space-less-ness would not exist. The zero matrix terms must equal the BRST charge and vanish in the negative shell and mathematically without fudge factors this can't be done so unless in a singularity space-less-ness doesn't exist with reference to conformal weight(conformal gravity)or conformal time with reference to space-time in the Bose Sea(Higgs Bosonic field).

CHAPTER EIGHT: BOUNDARIES AS IN CONVERGENTType equation here. FOURIER SERIES WITH ASYMPTOTES

The series of the infinite sum or

$\Sigma\ from\ n = 1\ to \infty\ of\ \frac{1}{2^n\,will}$ converge to zero as will the infinite product $\Pi\ n = 1\ to\infty\ of\ \frac{1}{2^n}$ which ties in with the spiral operator on space − time converging to zero as space − time constricts toward zero as the event horizon is approached.

DETERMINING THE STEADY STATE TEMP. OF THE UNIVERSE

The steady state temperature of 2 concentric circles(compactification of type IIa string theory with boundaries at temp $f\theta$ and $g(\theta)$ is as follows $u(r,\theta) = A0 + B\,0\ln r\ +$

$\Sigma\left(A\,n\ r^n + \frac{B^n}{r^n}\right)\cos n\theta\ +$

$\left(c\,n\,r^n + \frac{D^n}{r}n\right)\sin n\ \theta. A\,0 + B\,0\ln a = \int f(\theta)d\theta\ \left(\frac{1}{2\pi}\right).\quad A\,n\,a^n + Bn\,a^{-n} =$

$1/\pi \int_0^{2\pi} f(\theta)\cos\theta\ d\theta$ so $u(r,0) = A\,0 + b\,0\ln r + \Sigma\{A\,n\ r^n + B\frac{n}{r^n}\ n\cos\theta + C\,n\,r^n +$

$D\frac{n}{r^n}\,n\sin\theta$ with approaching zero density4 ... A 0=10^-24 cm B 0=10^10 light-years(approximate radius of this

universe)$\Sigma(1/\pi \int_0^{2\pi} f(\theta)\cos n\theta\ d\theta + 10^{10}lightyears\left(10 - 24\frac{cm}{10^{10}lig}htyears\right)^2.\Sigma n =$

1 to ∞ states $(-2n)times\ 10^{-24}/$
$10^{-10})^2. The conversion of lightyears to cm is aa huge number in the denominator

Denominator making the ambient temperature approach absolute zero as it is 2.74 degrees kelvin.

The velocity of space-time which is is accelerating either geometrically or arithmetically was determined as
$c\int_0^{2\pi} f(\theta)d\theta$ was determined to be −
$2\pi c$ from the Big bang to the point of maximum expansion back to the Big Bang

The equation specifically determined from the Fourier Series of an every expanding diameter never reaching a circumference of space-time where the diameter is + and − Riemannian Forces of nature we got
$1/\pi \int_0^{2\pi} F(\theta)\cos n(\theta)d\theta$ and since the trajecotry of The Big Bang is π radians or

180 degrees for an isotropic universe that is relatively homogeneous we
get$\frac{1}{\pi} f(\theta)d\theta \cos n\pi =$
$(-1)/$
$\pi \int f(\theta)d\theta$ or $1/$

$$\pi(-1)\int_0^{2\pi} F(\theta)d\theta =$$

$0 - (-2)c \; as \; the \; speed \; of \; light \; by \; the \; Riemann \; Forces \; after \; the \; Big \; Bang \; couldn't$

exceed "c" but space-time could. This results in (-1)-2πc n/π =2 c n where n is the infinite sum of all velocities from n=0 at pre-Planck Time to near infinite velocity at a Big Crunch or near the event horizon of black hole where space-time curvature approaches infinity constricting towards zero.

π by definition is the circumference divided by the diameter and the diameter is the sum of all Riemann Forces in terms of closed strings while the circumference refers to space-time. $\Sigma\, n = 1 \; to\infty(from \; flat \; space - time \; to \; infinite \; curvature \; space - time \; in \; a \; point) =$
$4\pi\rho \; from \; Poisson's Equation \; where \; the \; field \; is \; the \; Riemann \; Forces \; and$

space-time curvature metric known as gravity curving space-time where time is being curved by mass. Thusly the circumference /diameter becomes $2c/4\pi\rho$ and $\hbar =$
$h\frac{}{2\pi}which \; is \; multiplied \; by \; the \; 4\pi\rho \; in \; the \; denominator \; making \; \frac{2c(2\pi)}{4\rho}h \; or$

$4\pi c/4\rho h \; becomes \; \pi c/\rho h$

This means the velocity of space-time is
$\pi c \quad divided \; by \; the \; energy \; density \; of \; matter \; times \; Planck's Constant.$
Planck's constant(h) is 6.63 x10^-34 joule-seconds so the velocity of space-time is
approximately $\frac{\pi(6.63x10^{-34})c}{\rho}$ $which is \; accelerating \; toward \; near \; infinity.$ If the energy density of matter is in joules then the joules cancel out leaving
$\pi c(6.63x10^{34}) \; sec \; or \; 3x10^{8meters} \, / \sec \; \pi(6.63x10^{34})meters \; for \; the \; space -$
$time \; current \; tih \; a \; measurable \; universe \; which \; is \; 10^{10} \, lightyears \; in \; diameter$

meaning that space-time is "outrunning" this universe as 1 meter is 1.0570266x10^-16 light year and 10^10 light-years is 10^10(1.057x10^-16 meters/light-year)which is 1.057x10^26 meters as 10^10/10^-16=10^26 meters while space-time is rocketing along with its current at
$(3x10^8)6.63x10^34\pi \frac{meters}{sec} \; or \frac{\pi c}{h\rho} = 21(10^{42}) \; for \; the \; current \; of \; space -$
$time \; while \; this \; universe \; is \; 10^{26} meters.$

Now that we know that space-time curvature in the circle which is the compactification of IIA string theory(closed strings)and M theory we know how many revolutions of n describe space-time velocity in n(pi)c where space-time curvature is in radians of the circumference of the compactified circle. We also know that the curvature of this universe which is 5X10-29 radians is also approximately 3.3x10^5 meters of curvature intoto for this universe mostly concentrated about black holes.Of course this space-time curvature affects the velocity which is why

h

ρ must be incorporated and as photons have mass and light is bent by gravity h must

must be multiplied by the energy density of matter .The equation of everything can now also link the denominator of

Rab$\rightarrow \rho$ ab without using the approximation of k as $\frac{1}{c2}$. This is by using the relation

Rab=hρ ab or ρ ab = $R\frac{ab}{h}$ or $Rab = i\hbar\Lambda\ ba\rho$ ab $\mathbb{R}(n) = \frac{\prod(n=1\ to\infty)1}{2^n\pi R\infty}$ $+ 2\pi(Rabc +$

or $-\frac{1}{2}R\ g\ ab$ $\overline{R\ ab}$ note that the
2^nπ in the denominator of the spiral operator incorporates the
2π in the denominator of $\hbar = h/2\pi$

In the n=1 case the
$2\pi R\infty$ or $2\pi p\infty$ where $p\infty$ is the infinite momentum limit($\omega\infty$ or $R\infty\ \hbar$ brings the

denominator back to
$4\pi\rho$ which is Poisson's Equation where the spiral operator is the derivative

operator of the vector fields of space-time being constricted and inertial mass doing the curving into the event horizon of any black hole. So R(n)=the infinite product from n=1 to infinity of ½^n(pi)R infinity plus 2 π times [R abc + or -1/2 R g ab/ihΛ ba ρ ab]where Λ ba represents antigravity energy density for space and(R abc + or – ½ R g ab/R ab) is The Equation of Everything substituting R ab with ihΛ ρ (ab) in the denominator and 2π in the numerator. Of course this naturally includes any and all electromagnetic radiation by the equation
E(photon)=h
υ where υ is the frequency and is incorporated in the energy density of matter.

So to summarize $\mathbb{R}(n)abcd = \frac{\prod(n=1\ to\infty)1}{2^n}$ or $\prod 2^n + 1\ (\pi)\omega\ /2^n\pi\omega'$ $+ 2\pi(R\ a\ b\ c +$

or $-\frac{1}{2}R\ g\ a\ ab$ $\overline{R}$ ab which equals $\mathbb{R}(n)abcd = \prod\frac{2^n+1\omega}{2^n\pi\ \omega} + 2\pi\ [\ \Big(\ R\ abc +$

or $-\frac{1}{2}R\ g\ ab\ \otimes i h\Lambda\ ba\ \rho\ ab)^\wedge -$

1        $\Big)\ ]$ where ω is angular momentum from 1 to the infinite momentum limit

limit also known as p n where n =1 to n= infinity. Space-time approaches infinity without reaching it so R a b c+ or – ½ R g a b may approach infinity in an asymptotic flat space-time where R a b
c=$\mathbb{R}$ abcd if mass(R a b)approaches zero as in the case where the BMR is heating

Space on 2.74 degrees kelvin. Space-time is approaching infinity with eddys ,currents and backwash as galaxies rotate around central black holes that rotate differently due to differential hydrostatic pressure of space-time near the event horizon and the degree of electromagnetic radiation .So if total space-time approaches infinity then $\infty(6.63 \times 10^{34}) = \infty$ in the numerator and with the spiral operator as

n→

∞ *the operator approaches* $\frac{1}{\infty}$ *which when operating on a function that approaches*

ed the n dimensional state with infinite curvature constricts to a point but for the asymptotically flat first 3 dimensions of euclidian space with time curving around it as in Minkowski space in a near vacuum of almost massless space at 2.74 degrees kelvin the operator generates ½^n(pi)R where n=1,2 or 3. So here in deep massless space Riemannian space-time can approach infinity rather than zero as when constricted as Schwarzchild space-time yet when operating as a unit the spiral operator approaches anywhere between zero and infinity depending on the number of microscopic dimensions and degree of space-time constriction. Note the operator operates on a function or series of functions and in the case where the infinite momentum limit is applied the total Riemann constricts toward zero space-time without reaching it and in this case the spiral operator becomes zero operating on Planck's Constant(infinite Euclidian space with almost zero curvature to infinitely constricted space with infinite curvature)but in these cases the result is still zero.To achieve everything the inertial mass R ab or energy density of matter rho(a,b)must approach either infinite mass or infinite potenetial energy which can occur if the Bose Sea is the Higgs Boson or massive Higgs Field with almost infinite potential energy and mass yielding

$\frac{\infty}{\infty}$ *which is everything except zero and as zero is asymptotic it'snever reached.*

CHAPTER NINE: OUTLINE OF FLUID AND HYDRODYNAMICS AS APPLIED TO VELOCITY AND CURRENT OF SPACE-TIME WITH REFERENCE TO BLACK HOLES AND TO THE HUBBLE EXPANSION COEFFICIENT(H 0) FOR GALAXIES

H0 T0=

$H 0 \int_0^{a0} \frac{da}{a}$ which is the integral of $\frac{du}{u}$ for the spiral fractal formula for space --

-- time. This is $\int_1^{\infty} dy/y[\Omega y^3 + \Omega R y^2 + \Omega\Lambda]^1/2$ where $\Lambda =$

cosmologic constant and Ω relates to the density of this universe with reference to the expansion of the universe as it ties in with a 0 and y=1+z=a 0/a...5   H 0=100 h-km/sec Megaparsecs-1 and H 0^-1=Hubble time=0.98x10^10/hour-years with a universe whose expanse is 10^10 light-years. Hubble length=c/H 0=3000/h-Mega-parsecs and the critical

density=$3\Omega H^2 \frac{0}{8\pi G} = \Omega\rho$ critical $= \frac{2.78x10^{11}\Omega h^{2M}}{Megaparsec \ s^3} = 1.88 \frac{x10^{-29}\Omega h^2 gm}{cm^3}$ which relates to

the total space-time curvature of this universe of 5x10^-29 radians in a 10^54kg universe or 3300 k M or3.3x 10^5 meters. This indicates a relationship between space-time curvature and critical density . When the critical density is reached THE ACCELERATED EXPANSION FOREVER STOPS whereby as if in a supermassive black hole such as the one which marked the Big Bang if the critical density was exceeded by a factor of 2 to 2.5 this indicates by the space-time curvature metric(converted to radians) a collapse of another universe that underwent a Big Crunch occurred just over 10^-43 seconds prior to "The Big Bang" proving a massive implosion prior to "The Big Bang "had occurred .(Of course an alternative theory is the collision of two membranes triggered "The Big Bang" possibly an a 50:50 matter antimatter mix.) Note also if the critical density is exceeded space could break up and diffuse out into space-time adjacent to the area involved causing ripples or gravity waves adjacent to supermassive black holes. Also ,finally, this explains WHY space-time constricts as it approaches the event horizon of black holes ;if criticality is met as the event horizon is breeched and space actually breaks under the crushing pressure and super density at the event horizon the space-time curvature increases toward infinity and a mini-Big Crunch occurs at each and every black hole with counter-flow hydro-dyanamics of the perfect fluid of space-time spuming out with Hawking Radiation. Hydrodynamics can be applied to space-time as space-time satisfies the perfect fluid equation and this can be applied to black holes. The trajectory of Hawking Radiation from the event horizon can be considered similar to a radial flow pump.ω can be considered the radial velocity with r1 as the radius of the inner core of a spiral matrix leading to the convergence of Schwarzchild Space-time while r2 is the spiral outer core or shell of collapsing space-time with an outward tangential velocity or trajectory called V2 with (V2)r and V

θ2 as components of the tangential velocity whose angles areα2 andβ where α2 is is the angle between $v\theta$2 and V2 and β is the angle between$(v2)r$ and V2.The spiraling in is the control volume of space-time and the control volume is$\Sigma Tz = \rho Q(r2V2 \cos\alpha2 -$

$r_1V_1 \cos\alpha_1$). *This based on Torque free Euler's Equation* $r_2V_2 \cos\alpha_2 = r_1V_1 \cos\alpha_1$ *where the fluid spirals freely either radially inward or outward. torque*

FOOTNOTES;1.HARD EVIDENCE FOR THE MULTIVERSE FOUND,BUT STRING THEORY
www.mathcolumbia.edu/-woit/wordpreview/?p.5907

2.ibid.Pierpaoli,Elena et.al, Houghton, Richard,et.al, Mirsini-Hillman, Laura et.al

3.Kaku,Michio .Introduction to Superstrings and M Theory.p.165(Conformal Field Theory)

4.Spiegel,Murray. Fourier Analysis with Applications to Boundary Value Problems p.50

5.Peebles.Principles of Physical Cosmology.p.100-103 Princeton Press

6.Granger,Robert,Fluid Dynamics.p.276   7.Hau,Lene.Harvard Gazette .Physicist Slow Speed of Light

Space-time acts as a perfect fluid with a velocity of$(h^\wedge - 1)\pi c/\rho$ or near 2x10^35 (c) which explains the huge pull of space-time coupled with the push of Dark Energy to explain the accelerated expansion of this universe. Whether the pull or current is being accelerated by near infinite velocity for space-time due to a singularity or Big Crunch in another universe is still up to debate although there is some empirical evidence that a multiverse does exist. The ocean of space-time approaches infinite size with eddys, currents and backwash especially with drag caused by the trillions of black holes which curve space-time to 5X10^-29 radians(2π *radians is* 360 *degrees*) or approximately 3300 kilometers(3.3x10^3km) or 3.3x10^5 meters(330,000meters) while open flat space occupies a vast majority of this universe with almost mass-less-ness and an ambient temp of 2.74 degrees kelvin.

The dimension of time R(t)=+ or -1/2 R g ab/ρ *ab*   where R g ab is the space-time curvature metric known as the effect of gravity and
ρ *a b is the energy density of matter which is converted from R ab inertial*
Mass($\sim$)1/c^2 or multiplied by Planck's Constant(h)

or$\hbar = h\frac{'}{2\pi}$ *where with Poisson's Equation using the dual vector field as space —*
time and inertial mass causes R ab =
$\hbar\rho$ *ab   and ρ ab where ρ ab is the energy density of all the RiemannForcesof nature*

a nd the circumference is space-time in the compactified circle of type IIA string theory.
So technically R(t)=Λba + or – ½ R g ab/ $i\hbar\Lambda$ *ba*ρ *a b* + (+or $-\frac{1}{2}$R g ab $\frac{}{Rab}$ =
circumference of compactified circle. In this case
as 10^{-52} *in the numerator isIn this case*Λ *ba reflects an outward curving of space*
     — small albeit curving or flattening of space — time for anti
     — gravity so $+\frac{1}{2}$*R g ab would apply. The space*
     — time curvature metric known as the effect of $\frac{gravity}{}$*which*

is curved by the space-time curvature metric gives space-time for the Riemann Forces of nature in terms of the energy density of matter. In other words the Riemann Forces of nature are + or -1/2 R g
ab/$i\hbar\Lambda$ *ba ρ ab which are the space —*
time curvature metric divided by Planck's constant times the energy density of matter where iΛ bahρ a b = R ab or inertial mass. (i$\hbar\Lambda\rho$ ab = R ab)
Also note that gravity waves were detected with LIGO L shaped detectors with a laser interferometer gravitational wave observatory from two separate sites Washington ,D.C. and Louisiana and the two detectors measure almost immeasurably small length divergences using laser beams as well as around black holes ;coalescing and individual. The configuration of space-time is an ocean of near infinite length with whirlpools near the event horizon of each and every black hole which is active. These whirlpools form a

continuous drag on the accelerating expansion and as the momentum(p or R)approaches infinity where space-time constricts toward zero the sum total of all active black holes provides a significant drag on the expansion. When the black holes (dry up) space-time in that region approaches a steady state which no longer provides a drag on the acceleration.When two black holes coalesce into one supermassive black hole the spiraling in Schwarzchild Space-time coalesces into one huge spiraling area of space if the angle between the black holes is

2π *radians or* 360 *degrees and if the proximity changes to where the angle* reduce s toward 180 degrees or

π *radians the area of coalesence changes according to the cosine of the function*

N of the net amplitude of each component black hole. The cosine of 180 degrees orπ *radians is* −

1 *so the amplitude is reverse in amplitude to when the angle is* 0 *radians.*

As the cosine of 0 radians is 1 when the angle between the two coalescing black holes are additive while at

π *radians the net effect will be the amplitude of the stronger black hole or* 0
                                 if they are a

They are approximately the same amplitude .Of course two black holes attempting to coalesce that are almost equal but opposite in magnitude 180 degrees from each other will cause a probable singularity which may break up or rip space-time causing a Big Crunch.

THE AGE OF THE UNIVERSE;IS IT YOUNG OR OLD?

   The Big Bang was postulated as being 13.7 billion years ago. But does this mean that this universe is young or old? Edgar Alan Poe stated that when the universe is old the sky will be full of stars to the point where starlight will illuminate the sky rather than it being dark with only sporadic stars, planets and galaxies. This is because he believed that nebulae will continue to form stars from gasses such as hydrogen fusing into helium at a continuous rate which is greater than the rate stars collapse into dwarfs from red gas giants like Betelgeuse(very old star)or collapsing into black holes. The point is the rate of rotation of the expansion of the galaxies is decreasing geometrically over time as the expansion is increasing geometrically. If one could localize the black hole which locates the Big Bang; the rate of rotation near the Schwarzchild Space-time(even if the black hole isn't still active)can be used to determine how "old" this universe is at 13.7 billion years and compared to other universes(if our technology is sufficient to measure them and prove their existence)to formulate a graph of universes in the multiverse to determine if our universe is young middle age or old compared to other universes or manifolds.
   Will this universe end in "Heat Death" or a "Big Crunch"? This point was touched upon by this author's previous book "Megaphysics II;An Explanation of Nature". "Heat Death "will result if the acceleration of the expansion of galaxies reverses in a

continuous slow rate like tapping on the brakes of an automobile until the stars are so far apart and gravity plus anti-gravity diminish in their effects until the curvature of space-time whether curved inward with gravity's effect or outward with Dark Energy's anti-gravity reaches a constant rather than A DIMINUTION OF GRAVITY WITH AN INCREASE OF ANTI-GRAVITY OVER TIME resulting in an increase in flatness of space-time as the ambient temperature of all galaxies including stars approach 2.74 degrees kelvin which is incompatible with most electrochemical reactions as photon slow down in a matrix of Boso- Einsteinian Condensate(discovered by Dr.Lene Hau in 1995 at Cambridge,Mass.)This can be a monumental discovery as all attempts by physicists and chemists to reach absolute zero are in actuality reaching 2.74 degrees kelvin. Indeed, even physicists who have stated that there is a different kind of matter that goes to a thousandth of a degree below absolute zero are actually going to a thousandth of a degree below 2.74 degrees kelvin. Quantum Mechanics states that if the measuring device is part of what's being measured, the result will be skewed .Of course relative temperatures can be calculated incorporating the B.M.R. heat understanding that the figure is a steady-state figure for the B.M.R.Unless there is a way to totally eliminating the B.M.R from the Big Bang in an isolated area(which seems to be beyond our technology)it is extremely impossible to measure temepratures at zero degrees kelvin;only down to 2.74 degrees kelvin.AS A CONSEQUENCE IN THIS UNIVERSE THE REASON WHY ABSOLUTE ZERO CANNOT BE REACHED IN NATURE IS BECAUSE THE BACKGROUND MICROWAVE RADIATION FROM THE BIG BANG WILL HEAT UP EVERYTHING IN THIS UNIVERSE BY 2.74 DEGREES KELVIN AND IT SEEMS WITH OUR TECHNOLOGY IT WILL BE EXTREMELY DIFFICULT TO GO BELOW 2.74 DEGREES KELVIN AS ONE CANNOT PURGE ANY GIVEN AREA FROM THE BACKGROUND MICROWAVE RADIATION FROM THE BIG BANG!!This is reason why absolute zero cannot be reached in nature.Believe it or not,this skew correction may be affecting thermometers thoughout the planet as they may be off from the actual temperature by 2.74 degrees kelvin or centigrade.(1 degree kelvin=1 degree centigrade) F degrees =9/5 c+32 so (1.8)(2.74)+32 may be a the skew measurement of Farranheit thermometers unless they have it factored into the measurement of the temperature already(already measuring the heat of the B.M.R. plus the environmental temperature).Despite this unless the 2.74 degree correction has been made in experiments to hit absolute zero the temperature readings in the laboratory will be too low by 2.74 degrees kelvin.

If a "Big Crunch" occurs from a geometric rather than arithmetic acceleration of the galaxies everything collapses in upon itself like forming a huge supermassive black hole either in a slow phase (space-time's velocity will have a reduction in it's acceleration with a concomitant start of a dilation of time and a shift away from the infrared toward the ultraviolet)followed by a rapid phase which can occur in Planck Time 10^{-43} second once a threshold is reached which may be the critical densityΩ *critical or ρ critical. Despite this* if this universe is still very young compared to other universes(if they exist)then "Heat Death"would be far off in the future as the night sky is like a photograph of a bullet being fired where the firing is The Big Bang as a "stop action photograph"of the "movie"of the expansion-rotation,expansion,accelerated expansion like a bullet just after it was fired before inertial mass slows the bullet down until it is finally stopped by resistance of "the other universes if they exist "unless of course space-time has mass in and of itself which is only up to scientific speculation at this time. If space-time can show resistance as a perfect fluid with counter-currents does this indicate that space-time has mass .It is the opinion of this author that if space-time had a mass it would show almost equal but opposite reciprocal curvature to the curvature caused by the inertial mass of the contents of the container. This would on the surface show totally flat space-time which it isn't based on the measurement and detection of gravity waves. What may be considered a simple concept would in all actuality be immeasurably complicated indicating that space-time would have momentum which may go along with currents. Query;to be a perfect fluid does the perfect fluid have to have mass.Answer:if space-time has mass it would have to occupy a state of matter.If satisfying the perfect fluid equation and behavior like a fluid indicate that space-time is in all actuality a fluid. If this was true then at or around < 2.74 degrees kelvin space-time would congeal to Boso-Einsteinian Condensate and possibly solidify at absolute zero just as matter does. So using logic as an axiom space-time exists,corollary space-time acts like a perfect fluid, corollary space-time constricts as it approaches the event horizon of each black hole,corollary space-time can express a "pull" or" push". Are these conditions necessary or sufficient for space-time to be considered to have mass? If necessary space-time can still be massless;if sufficient space-time has mass. Light is bent by gravity and sucked into black holes as space-time constricts .Adjacent to this region space-time is less constricted and approaches asymptotic flatness .If space-time has mass then the Boltzman Equation for the states of matter would apply to space-time. AXIOM: Nothing is massless Corollary: Nothing has no properties. True or false If mass-less-ness is a property then nothing has properties when by definition nothing has no properties. The axiom and corollaries for space-time are NECESSARY for space-time to have mass as it would be for electromagnetic radiation such as photons to have mass, but are they sufficient for space-time to have mass?Photons have a non-resting mass which was measured at $\sim 3 \times 10^{-18} eV/c^2$ per photon but there are absolutely no measurements of a mass for space although the infinite momentum limit may include the constriction of space by the dilation of time at the event horizon of any active black hole.

   Does mass create space or just require space? Mass cannot exist without space(axiom)but is the converse true?If space or space-time has mass then the mass of that unit of space-time is the space of that space-time.That is true. Axiom :There is a law of conservation of energy. Corollary :Energy and mass are inter-changible and express entanglement .Corollary:A Law of Conservation of mass has been postulated just as a Law of Conservation of Momentum has been postulated.If mass created space then space could be created or destroyed.FALLACY!!!There is a law of conservation of dimensions in each and every system even if space is constricted by time ;the n dimensional state still exists although in a constricted state these dimensions still exist even if string sized 10^{-33} cm or smaller in dimensional states which are so microscopic that they can be infinitely small as part of a super-BRANE which is so submicroscopic that they hold together all other states which are below Planck Length and under Hilbert Space. There may be a lower limit for BRANES coupled with multiple energy states(converted from matter as part of entanglement as energy density of matter) but as a multitude of non-parallel planes from the subdivision of space-time into smaller and smaller slices in motion and curved there are an infinite number of intersections. Albert Einstein coined the term "space-time continuum"meaning without beginning or end.This statement implies that space-time is infinite and therefore space-time cannot be created or destroyed.

Is an intelligent observer required for space-time to exist? This question gets into philosophy ,religion and quantum mechanics as well as the weak Anthropic Principle. As reality seems to be altered by the presence of an observer is an observer necessary or sufficient for reality to exist? This gets into the question of whether or not electrons including spin and position is altered by observation. When the object being observed is part of what's being measured then the measurement is being skewed by the observer. As nothing doesn't exist and space-time does ,does that mean that there is always an intelligent observer .Also is nothing doesn't exist then no observer would exist to observe nothing as the observer is part of what's being observed.If a tree falls in a forest and nobody is there to observe it does it make a sound? Based on classical physics the answer would be yes ,but quantum mechanics puts a doubt on that answer indicating that reality is altered by the presence or absence of an intelligent observer. Objective evidence to answer this question is wanting eventhough it has been demonstrated that the presence of an observer does effect the properties of elementary particles; though in general this axiom seems to defy logic.

The Equation of Everything can be utilized to elucidate whether or not space-time has mass.$R(n)abcd = \Pi 1/(2^{n)pi} R$ $[+ [(R \; abc - or + \frac{1}{2} R \; g \frac{ab}{R} a \; b]$. $LET \; R \; ab =$ 0 $as \; inertial \; mass \; of \; space - time. Then \; R(n) = \frac{\Pi 1}{2^{nR} \infty} [\Lambda ba + (R \; abc + or -$ $\frac{1}{2} R \; g \; ab \; / \quad R \; ab \; where \; R \; ab = 0 \; so \; R(n) =$ $\infty \; so \; Riemannian \; or \; Lorenzian \; Curved \; space - time \; is \; \rightarrow \infty \; and \; if \; Rabc + or -$ $\frac{1}{2} R \; g \; ab = 0 \; then \frac{0}{0} is \; everything \; but \; as \; zero \; gravity \; is \; impossible \; R \; g \; ab \neq$ 0 $which \; means \; mass \; mass \; or \; its \; energy \; equivalent \; must \; always \; exist \; and \; as$

0 space-time is space-less-ness is impossible the numerator can never be 0 so the equation is the spiral operator operating on $\omega(n) + 2\pi R \; abc + or - \frac{1}{2} R \; g \; ab/0$ where n=R abc- or +1/2 R g ab≠ 0 $therefore \; space - time(Riemannian \; n \; dimensional)is \; \infty \; which \; is \; true \; space -$ $time \; is \; infinite \; or \; a \; continuum \; as \; \epsilon \; or \; a \; small \; number \; for \; the \; for \; inertial \; masss$

inertial mass(R ab) for space-time would bound it as finite if it's has a small mass which isn't 0 but approaching 0. Of course $\Lambda \; ba \; is \; the \; cosmologic \; constant \; representing \; the \; energy \; density \; of \; nearempty \; spacesssp \; s$ space and the ba in this case works as anti-gravity to inertial mass R ab.However the energy density of almost massless space-time from the BMR is due to the mass of photons and reflected as anti-gravitational as a push reflecting Dark Energy .This is why the cosmologic constant is so small a number .The matter of the BMR is only due to the mass of photons and is inversely proportional to the square of the tensor of space-time(not the curvature metric or variant)In black holes however, space-time appears to have a "false mass" .This is possible due to momentum , angular momentum and torques involved with the transfer of energy states from ordinary matter antimatter and photons primarily involved in singularities like a Big Crunch or active supermassive black holes which distort and constrict space-time due to the high density of the black hole curving space-time toward infinite curvature where mass would gravitate and possibly diffuse into space breaking it up as it super-constricts .Again the infinite momentum limit of the spiral operator in the denominator isn't important as $0\infty = 0$ *and the operator is separate from the function being operated on.*

CONCLUSION :In a non-singularity space-time is massless but absolute mass-less-ness is impossible as energy such as electromagnetic radiation is composed of photons which have mass so in a singularity such as a Big Crunch or black hole space-time has a mass small or miniscule but possibly greater as space-time curvature increases toward infinity as constriction forces space-time toward zero. This is why space-time spirals into black holes in a constricting manner as Schwarzchild Space-time in an ocean of almost massless space-time acting as a perfect fluid or near perfect fluid.

The Cosmologic Constant and its relationship to space-time.The stress energy tensor $Tij=\Lambda\, g\,\frac{ij}{8\pi G}$ *is the cosmologic constant of the metric gij* $\div$
$8\pi G$ where G is the gravitational constant $6.67 x 10^{-11}$ newton meters/sec^2. Also$\Lambda\, g\, ij =$
$\frac{1}{R^2}$ *as an approximation where R ij is the space* $-$
time expansion indicating that $\Lambda\, g\, ij$ is a very small number with respect to the metric g

Metric g ij where i=initial event and j=final event and the stress energy tensor T ij is stress energy of the metric g ij with respect to inertial mass and the space-time curvature metric of gravity. Now we know from the action formula S=-1/2k^2(-g)^1/2 R where R is the space-time curvature variant of the metric g and s is the most likely action of that metric with k being the gravitational coupling constant that if the metric were $\Lambda\, g\, ij$ *then the action would be* $S = -\frac{1}{2k2}$ Λgij $\frac{(-g\, ij)^{1/2}}{}$ $R\ ij$ which relates R ij
^2$\propto \frac{1}{g}$ ij $\propto$
$\frac{1}{\Lambda}g\, ij.$ *Here the space* $-$
tie curvature variant R ij relates to the square root of the inverse of the metric $-$
g ij by inserting the cosmologic constant with reference to the metric g ij and the inverse

The inverse of -1/2 the gravitational coupling constant squared to bring the most likely action of the metric g ij with the cosmologic constant factored into the gravitational coupling constant for g ij.

$$\Lambda\, ij = \frac{1}{Rij^2} = 8\pi G\ Tij$$

The inverse of the metric g ij relates to the space-time curvature variant and the inverse of the cosmologic constant with reference to the metric g ij relates to the square of the space-time curvature variant which equals the stress energy tensor times 8
$8\pi G$. *Generally the cosmologic constant with reference to g ij is inversely proportional to the sq*

tional to the square of the space-time curvature variant which equals the stress energy tensor times $8\pi G$ *where G = the gravitational constant.*

The cosmologic constant Λ is the value of the energy density of the near vacuum of space(temp 2.74 degrees kelvin from Background Microwave Radiation from The Big Bang)from Albert Einstein in 1917.So actually The Equation of Everything in terms of Stress Energy is $\mathbb{R}(n) = \Pi\, n = 1\frac{to\,\infty\,1}{2^n\,\pi Rn}$ $\quad \hbar 8\pi G Tab(R\ abc + or - \frac{1}{2}R\ g\ ab \div$
$R\ a\ b$ *where $8\pi G\ T\ ab =$*
$\Lambda\ g\ ab$ *where $T\ ab$ is the stress energy tensor of matter and Λ is the energy density o* density of space with the
B.M.R.$8\pi T\ ab = R\ ab - \frac{1}{2}R\ g\ ab$ which locates the Einstein Tensor $G\ ab$. So $\mathbb{R}(n) =$
$\frac{\Pi\ 1}{2^{n\pi p}} \hbar G(R\ ab + or - \frac{1}{2}R\ g\ ab \left(R\ abc + or - \frac{1}{2}R\ g\ ab \div R\ ab\right)$ *where $p =$*
momentum and ω is angular momentum. Here the $\frac{1}{2}R\ g\ ab's$

cancel leaving $R\ a\ bc \frac{}{R}ab$ times$\frac{R}{R}ab = Ra\ b\ c\ \div R\ ab$ times $1 = R\ a\ bc\quad \div$

$R\ a\ b$ =R(n)abcd or Riemann space-time=flat Euclidian Space/Inertial mass as indicated by the Ricci Tensor which of course is The Equation of Everything   DARK ENERGY AND IT'S EFFECT ON SPACE-TIME
Dark Energy has a flattening effect on space-time .Dark Energy and the cosmologic constant both illustrate anti-gravitational effects pushing galaxies away from each other .As space-time is curved OUTWARD in an accelerating expansion and Dark Energy from "The Big Bang" it has a flattening effect. Black Holes curve space-time inward toward an infinite curvature constricted region(Schwarzchild narrowing).The cosmologic constant or the energy density of space with an inertial mass~0 but non resting mass for the B.M.R. of 3x10^-18kV/c^2 per photon may be the ratio composed of approximately 68.3% dark energy and 26.8% dark matter with the remaining 4.9%being composed of atoms; where the cosmologic constant may be dark energy itself with it's anti-gravitational properties as a vast majority of this open asymptotically flat expanding universe is empty space with only the photons of the B.M.R. heating it to 2.74 degrees

kelvin.As the energy density of open flat expanding space is The CosmologicConstant as indicated by the lambda-CDM model of physical cosmology. The equation of Relativistic Gravity is R ab-1/2R g

ab+Λ g ab =

$\frac{8\pi G}{c^4}$ T ab so the $equation$ of $everything$ $incorporates$ $dark$ $energy$ as the

$$cosmologic\ constant$$

Cosmologic constant as R(n)=R a b c d=R a bc+or –R g ab

$+\Lambda\ g\ ab$ $\otimes$ $R\ ab^{-1} where\ n\ dimensional\ Riemannian\ Space-time=$
$Flat\ euclidian\ space\ plus\ or\ minus\ the\ space-$
$time\ curvature\ metric\ known\ as\ the\ effect\ of\ gravity+$
$the\ cosmolgic\ constant\ effect\ of\ anti-$
$gravity\ believed\ to\ be\ Dark\ Energy\ with\ the\ vector\ or\ tensor\ product\ of\ the$
$$of\ the\ reciprocal$$

the reciprocal of the inertial mass represented here as the Ricci Tensor but may be transformed into$\rho\ ab\ the\ energy\ density\ of\ matter+\Lambda\ g\ ab\ as\ the\ energy\ density$ of space-time with BMR.So the value of

$8\frac{\pi G}{c^4}$ $T\ ab\ is\ the\ actual\ solution\ to\ the\ equation\ of\ relativistic\ gravity.$

so

$\mathbb{R}(n)=\prod n=$

$1\ to\ \infty\ of\ the\ spiral\ operator\ \frac{1}{2^{n\pi\mathbb{p}}}\ acting\ on\ the\ reciprocal\ of\ momentum.$

$8\frac{\pi G}{c4}$ $T\ ab$ $\frac{\left(R\ a\ b\ c+\frac{1}{2}R\ g\ ab+\Lambda\ g\ ab\right)\otimes\ 1}{R}ab=$

$\mathbb{R}a\ b\ c\ d$ $where\ T\ ab\ is\ the\ stress\ energy\ tensor\ and\ \frac{8\pi G}{c^4 relates}to\Lambda\ with\ R\ ab\ as\ inertia$

inertia and +1/2 R g ab being the curving OUTWARD OF SPACE-TIME DUE TO THE EXPANSION FROM THE ENERGY DENSITY OF DARK ENERGY AS IDCATED BY THE COSMOLOGIC CONSTANT.

APPENDIX

Note also that the spiral operator

$\prod_1^\infty\frac{1}{2^n}$ $operates\ on\ \frac{1}{\pi Rn}where\ Rn\ is\ the\ momentum\ function\ with\ the$
$$infinite\ momentum\ limit\ as\ R$$

momentum limit where R is actually

$\mathbb{p}\ or\omega. This\ function\ and\ operator\ in\ The\ Equation\ of\ Everything\ in\ and\ of\ itself$

Proves space-time constriction at or near the event horizon of every active black hole.As this operator/function results in zero(0) this is multiplied to the tensor equation of everything R a b c+1/2 R g ab+Λ g ab $\otimes \frac{1}{R}$ ab to yield the Riemann Tensor ℝ a b c d as applied to n dimensions

with reference to the spiral operator. $\mathbb{R}(n) = \prod_1^\infty 1/2^{\wedge} n\pi$ operating on the reciprocal of the infinite Momentum Limit R∞ which equals zero(0) yielding R a b c d=(0)(R a b c+ or − R g ab+Λ g ab) $\otimes \frac{1}{R}$ ab. The numerical value of the operator operating on the variable of momentum either Rn

or ℙ n becomes an infinite product of momenta which can be considered

a series of constants which are multiplied to the tensor equation R Rabc+ or − ½ R g ab+

Λ g ab $\otimes$ R ab^{-1}; so the euqation of everything takes many different forms with

with different constants

$\hbar, \frac{8\pi G}{c^4}$, Λ g ab or Λ g ij depending on the reciprocal of the vector product

and whether or not ρ ab is used with or without Λ g ab in concert with R ab or not.

THE CONVENTION
$8\pi G = $
1 can be used to describe energy density using the term cosmologic constant.

The value of the Cosmologic Constant is $\sim \frac{1(10^{-52})}{meter\ s^2}$ or $\sim 3(10^{-122})$ Planck Units with Ω Λ

$= 0.6911 + or − 0.0062$ $H\ 0 = 67.74 + or − \frac{\frac{kM}{sec}}{Megaparsecs} =$

$2.195 + or −$
$\frac{0.015(10^{-18})}{sec}$. Dark Energy results from a negative pressure driving th the

accelerated expansion from a positive energy density. Λ > 0 or ρ > 0. The critical density of this universe is that density that stops the universe

from accelerating an expansion forever.
ΩΛ is the fraction of the energy of the universe due to the cosmologic constant

and describes Dark Energy this is in a universe where space-time curvature=0 or is asymptotically flat.6 This is why space-time curvature is only 5x10-29 radians intoto where

2π *radians is* 360 *degrees and distance wise only* 3300 *kilometers mostly near*

active black holes. Recall also that the cosmologic constant is inversely proportional to the square of space-time $1/R^2$.

7:Wikipedia;public domain

MISCELLANEOUS:CHAPTER TWELVE
The tensor characterisitcs of coherent vortex structures. WHAT DOES THIS
MEAN?Assume a vortex has two disparate component fields in
solenoids.
ω' and $\omega\gamma$If the field described byω'whereω has a chosen volume andω is outisdeω

De
outside ω' there is a gradient of the harmonic function inside the volume. For all
  ω' and$\omega\gamma$ satisfied if at boundary of regions the derivative of the function is
equal to the normal projection of the vorticies. There is an extant mathematical proof
using the Tensor Virial Theorem vortex-bulge systems using the Faber-Jackson Relation
$Rea\sigma(0)^2$where by $MH - \sigma\, 0 \rightarrow$
$M\,Ha\sigma\,0^4$. Considering spontaneous breaking of symmetry based on the spiral
fractal and spiral geometry for space-time as defined and discussed by Albert Einstein
in 1912 and mentioned in this author's first book ""Megaphysics ;A New Look at the
Universe" as well as being in this author's spiral operator for space-time it can be
shown that voticies of an abelian Higgs Field(characteristics of each subgroup may or
may not cross over into other subgroups)and divulge anti-symmetry with the vorticies
of the Higgs Field(can be equal and opposite so covariant and contra-variant tensors
can net out).This may possibly indicate a clockwise and counterclockwise rotation to
spiral space-time as indicated in this author's first book showing asymptotic flatness
and still having mathematical spiral geometry.

The anti-symmetric tensor field of spiral space-time interacts with the massive Higgs
Field(God particle????) via a Lagrange or LeGendre Transformation and using what is
termed the Kalb Raymond Action for strings; the abelian Higgs Model is postulated for
abelian Higgs Vorticies. Basically anti-symmetry in tensors is$(\Gamma.\nabla x)\Gamma = -(\Gamma.\nabla x)\Gamma$ so
that they're equal but oppsosite and therefore anti-symmetric. Of course abelian is
indicated by sharing or intermixing subgroups as
in$(\Gamma.\nabla x)\Gamma = (\nabla x.\,\Gamma)\Gamma$. These data are from The Journal of Mathematics(anti $-$
symmetric and abelian tensor fields). $\nabla$is the harmonic funtion$\Gamma$is the index
of sphericalBased on these findings
Spherical boundary. As a result of this boundaries of harmonic tensor fields relate to
space-time vorticies and possibly the Higgs Field.
Anti-symmetric or skew symmetric tensors are involved in Riemannian Space-time:
A^T=-A   and a ij=-a ji for all i,j where i is the initial and j the final event. Based on the
LaGrange Identity (r1xr2)^2=(r1^2)(r2^2)-(r1r2)^2(which had previously been used
in a previous book on the subject of the osculating and tangential plane by this author)
g=det(g
ij)>

$0. \left(\frac{ds}{dt}\right)^2$ *relates to space −*

time coordinates as in the line element and relates as $ds^2 = \frac{utilizing\, ds}{dt} \cdot g\ ij\ dx^{idy}\ j =$
$l^j \leq i$. In this way space-time manifolds such as Riemannian, Lorenzian or Minkowski
spacetime can be more easily translated into tensor fields so that ds^2=g ij dx^I dy^j
and the identity (X+y)y'+x-y=0.Recall that g ij=rirj and also tht g

$ij\mu^{i\mu^j}. R(n)abcd$ *which is* $n − dimensional\ Riemannian\ space −$
time as abelian and antisymmetric with regard to tensor
  fields $abcd, dcba, , cdba, cdab, dcab$ *etc in covariant and contravarient form show*
Based on these equations space-time follows spiral geometry with chiral components
which may relate   to Schwarzchild Space-time space -time at both extremes of the
event horizon as well as the Higgs Field. The gradient volume inside a harmonic
function(nabla) approaches
zero.. $\nabla x \to 0\ so(\Gamma.\nabla x)\Gamma \to 0 and\ with\ anti − symmetric\ properties\ as(\Gamma.\nabla x)\Gamma \to$
$0 then − (\Gamma.\nabla x)\Gamma \to 0. If − (\Gamma.\nabla x)\Gamma = −g\ ji\ and(\Gamma.\nabla x) =$
$g\ ij$ *then an abelian antisymmetric tensor field* $(\Gamma.\nabla x)\Gamma = (\nabla x.\Gamma)\Gamma and\ g\ ij =$
$g\ ji. If\ g\ ij = rirj\ and\ g = \det(g\ ij)\ then\ (g\ ij))^2 = g\ i^2 j^2 − (g\ ij)\ true\ if\ g\ ij =$
volume of the vortexa vortex as $((\Gamma\nabla.x)\Gamma \to 0$ *and* −
$(\Gamma\nabla.x)\Gamma \to 0$ *then it can be postulated that any vortex including space −*
time of a harmonic function $\nabla x\ where \Gamma$(Christoffel Symbol) =
index of the spherical boundary or interface as it approaches zero volume then the actio
then the action of the vortex field on the Higgs Field approaches the action for closed
strings as in the type IIa string theory which compactifies to a circle of decreasing
diameter from infinity towards zero volume or space-time for closed harmonic strings
with for an abelian anti-symmetric vortex field acting on the Higgs Field may
presuppose that the Higgs Field is acted upon by space-time and since space-less-ness is
impossible the Higgs Field may or may not require space-time but if space-time
requires the Higgs Boson or Field then this could be presumptive evidence using logic
and math that the mysterious point particle "r"mentioned earlier which is required for
space at time t could relate directly or indirectly with the Higgs Field although there is
speculation that the tachyon may also be involved. Generally all matter and it's energy
equivalent require space and bosons(as well as tachyons) appear to relate to gravity as
an effect so generally bosons and tachyons(see this author's first book Megaphysics;A
New Look at the Universe (2003)p.58-59)should require space-time to exist.Space and
time can neither be created not destroyed as even with the infinite planes of D-0-Branes
these D-0-Branes are space.Space is a container,and the vacuum involved in the D-0-
Branes is an infinite space vacuum as space-less-ness is impossible.Time had no
beginning although if the centrifuge effect from bosons and tachyons was the first event
forming the universe of strings and the multi-verse;time is actually a super-set
containing events.Even without the first event the super-set still exists and must exist in
order for events to occur within that set.Mass doesn't create space and events don't
create time as mass can't exist without space and events can't exist without time.Note
the members of the super-set of time as events are Abelian(each event in the superset is
related to each other event in the superset all related by the superset of time.)As a
consequence of this the massive Higgs Boson at ~10^77joules smeared along a massive

Higgs Field initially peppered with tachyons(per this author's first book)and near infinite potential energy requires SPACE AND TIME TO EXIST.Mass requires space and space-time while being constantly reformulated as a perfect fluid cannot be created or destroyed.Pleae note that using the spiral operator for the zero dimensional state comprised of an infinite number of parallel planes and a combination of tachyons and bosons because they converted to string universe(s)comprised of an a near infinite number of string dimensions and macroscopic universe(s)comprised of four dimensional space-time reveals the mathematical expression of 1/3.14159728 or pi. This mathematical conclusion indicates that 7/22 existed with the infinite parallel planes. This may or may not be a meaning calculation in an Ontologic Proof and as "religion"is not the intent of this publication this author cannot make any specific conclusions regarding "religion" in this text.However,the reader may or may not draw his or her conclusions regarding this mathematical conclusion. There is mathematical evidence whih can possibly explain multiple religions hoever,this is not the intention of this text. The Law of Conservation of Energy implies that mass can only be converted into energy or back into mass so the is theoretically also a Law of Conservation of Mass which includes the concept of mass-energy entanglement and explains to a certain degree Spooky Action at a Distance .Even all the dimensions in any one system must be a constant.With SpecialRelativity using the Lorenzian Transformations as a mass approaches the speed of light boundary it's length shortens;it doesn't disappear.This does not mean that width,height ot time cease to exist.Time dilates or lengthens relative to the observer but as mentioned before it doesn't stop;it's an asymptotic function.The same goes with length,width and height.They still exist but are distorted with regards to measurement .As a result while in theory time=infinity indicates it goes infinitely slow or appears to stop it doesn't stop but approaches stopping without reach it being asymptotic.The same goes with length,width and height.Space-time constricts toward zero but never reaches it as infinitely dialted time wraps around space constricting it down toward an infinite curvature point but the point doesn't disappear,ever!Note according to Plato the"God field" has two components; one external to space the other interior to space however this would imply that the exterior would be time independent and the interior time dependent.This is actually a loose fallacy as shown earlier anything with mass is dependent on time as long as events occur and it requires space as a container. Despite this tachyons can act time independent possibly able to travel backwards in time (reversed time arrow) .As time as direct function of the space-time continuum it should be immutable (not created or destroyed, and consequently the Higgs Boson+Tachyon/Field should be inclusive(with everything)including the space-time continuum but if sufficient is it also necessary? In this authors opinion it is necessary This can get into philosophy and even religion which isn't the which the objective of this text; however if space-less-ness in a multiverse doesn't exist is space-less-ness sufficient for the Higgs Field along with space-time? In other words does the Higgs Field require space? According to Plato the answer that as being "impossible" (beyond te comprehension of mankind)existence can not exist in non-existence because if it exists there is no non-existence.But Quantum Mechanics says every possibility can and will happen but is nothing, non-existence and space-less-ness excluded? Yes because nothing is the only {} not inclusive in everything. The null set includes nothing and any postulated Higgs Vortex Field or Boson or tachyon are by definition not nothing and nothing doesn't exist although in math there are "mathematical ghosts" to refer to missing factors in math to complete essentially incomplete calculations such as space-less-ness but they can be considered as "fudge" factors but is the square root of -1 or i a mathematical ghost?Doubt i as the imaginary number has clear mathematical definitions and ghosts are discouraged in coherent mathematical proofs. Consciousness waves have been postulated(K-Suryon Waves)which are 1.6 times Planck Length and they have an associated wavelength and mass. The mass is 4.6x10^-64kg and the wavelength is approximately 7.647x10^20 mts. which is very low frequency and very low energy ;lower than radio waves. The mass of a K-Suryon Wave is designated as "d".This can be applied to the Ontologic Proofs to demonstrate senscence in "The Higgs Field which incorporates Tachyons(Megaphysics ;A New Look at the Universe)where this author described a matrix of a Bosonic Field or Higgs Field with tachyons. The consciousness wave would be V

$cw/\omega = Vcw^3 d^2 = 1.534x10^{-103}$ where $\omega =$

$\frac{mc^2}{\hbar}$ all called the Sina Equation of Consciousness. 9This is according to Bioomed

GODEL'S ONTOLOGICAL PROOF

Note also, Kurt Godel had a proof for the existence of "The God Particle".Godel's Ontological Proof8. Ax:1{Pφ) $\wedge\Box$ $\forall[\varphi(x) \rightarrow \psi(x)]$} $\rightarrow P(\psi)$

Axiom 2:P($\neg\varphi$) $\leftrightarrow \neg P(\varphi)$ Theorem 1:P(φ) $\rightarrow diamond\exists x[\varphi(x)$}

Definition 1: G(x)$\Leftrightarrow \forall\varphi[P(\varphi) \rightarrow \varphi(x)]$

Axiom 3:P(G)   THEOREM 2:DIAMOND OPEN$\exists XG(x)$

Definition 2:φ ess x $\Leftrightarrow \varphi(x) \wedge \forall\psi\{\psi(x) \rightarrow\Box \forall y[\varphi(y) \rightarrow \psi(y)]$

Axiom 4: P(φ) $\rightarrow\Box P(\varphi)$

Theorem 3: G(x)$\rightarrow G.ess X$

Definition 3:E(x)$\Leftrightarrow \forall\varphi[\varphi$ ess x $\rightarrow\Box \exists y\varphi(y)]$

Axiom 5:P(E)

Theorem 4:$\Box \exists x G(x)$

Ess=essential.an open diamond represents a a nabla with a delta under it without the central parititon or an open diamond.The square areas are supposed to be empty squares.

8:WIKIPEDIA:PUBLIC DOMAIN "Godel's Ontologic Proof"

Spacetime ds^2=dx^2+dy^2+dz^2-c^2dt^2+dr2=R abcd=$\Pi\frac{1}{2^{nπpn}} + \hbar + [\Lambda$ ab $+$ $R a b c + \frac{1}{2}R g \frac{ab}{Rab}$]or $\Pi\frac{1}{2^{nπpn}} + \hbar + [\Lambda$ ba $- \frac{1}{2}R g \frac{ab}{R}$ ab] andsub in case of $- \frac{1}{2}R g$ ab andΛ ab in case of $+ \frac{1}{2}R g$ ab we get $\frac{1}{Rg}$ ab and$\frac{1}{R}g$ ba which are are anti $-$ symmetric and abelian as are $\frac{1}{R}g$ cd and and$\frac{1}{R}g$ dc which is a very small number

So flat R abcd is a large number and R(t)=R d=-R d is a very large number.as a result that time R (t)curves space(Euclidian) the more the space is curved by time or R(t) or R^d constricting infinitely large space down towards a point with maximum constriction and infinite time curvature.Note that that the cosmologic constant flattens space-time by the inverse square or if R=space-time curvature

$\Lambda = \frac{1}{R^2 so}$ if $\frac{1}{R^2 is}$ substituted for Λ in the equation of everything then $R\,abcd \to k +$
$[R\,abc + or - \frac{1}{2}R\,g\,ab + 1/R^{\wedge}2]$ /R ab in a flattening expanding universe where R abc is flat Euclidian space R abcd is curved Lorenzian space-time, R g ab is the space-time curvature metric causing the effect of gravity, R is space-time curvature and R ab is inertial

mass. $k = \prod_1^\infty \frac{1}{2^n}\pi\mathbb{p}n + 2\pi[Rabc + \frac{1}{2R}g\,\frac{ab}{h\Lambda i\rho}$ and if $R \to$

∞ as in a point $\frac{1}{R^2 approac}$ approaches zero which is the cosmologic constant in a black hole where the extreme gravity and mass nullify it. In a near vacuum of space R abc$\to \mathbb{R}\,abcd(n)$ where $R(curvature) \to 0; \frac{1}{2}R\,g\,ab \to 0; R\,ab \to 0$ and $\Lambda \sim 10^{-110} \to 0$ where R abc$=\infty = \mathbb{R}abcd.$ This mimicks open flat expanding space where anti $-$ gravity is predomiant. R the space $-$
time curvature is reciprocal curvature to gravity so space $-$
time curves outward rather than inward than inward flattening space

Making
$R \to \infty$ and $R^2 \to$
∞ with reciprocal curvature which is negative curvature pushing and
pulling outward causing flattness
outwardly causing flatness from anti-gravity expanding space while in a black hole predominated by gravity space-time is constricted toward a point of infinite POSITIVE CURVATURE INWARD .If 1/R^2 becomes in

Type equation here. $1/-\infty^2$ with reciprocal curvature and antigravity $\frac{1}{\infty} \to 0$ for
COSMOLOGIC CONSTANT. Note the reciprocal curvature of space-time is $-\infty$ and $-$
$\infty^2 = \infty$ so the cosmologic constant is $\frac{approximately\; \frac{1}{1}}{\infty}$ or $0;$ BIBLIOGRAPHY   1.HARD

EVIDENCE FOR THE MULTIVERSE FOUND,BUT STRING THEORY DOESN'T..www.mathcolumbia.edu/-woit/wordpreview/?p/5907
2.ibid.Pierpaoil, Elena et.al.,Houghton, Richard et.al,Mirsini-Hillman,Laura et.al
3.Kaku,Michio.Introduction to Superstrings and M Theory.Springer Press.1998
4.Granger,Robert A..Fluid Mechanics.Dover Press 1995
5.Spiegel,Murray.Fourier Analysis with Application to Boundary Value Problems.McGraw-Hill Schaum Outline 1974
6.Peebles,P.J.E.Principles of Physical Cosmology.Princeton University Press 1993
7.Wikipedia.Public Domain" The Cosmologic Constant"
8.Wikipedia .Public Domain" Godel's Ontologic Proof"9:Kodukola,Siva Prasad. NeuroQuantology and Consciousness Biomedical Science and Engineering 2014.p.4852 Kodukola,Siva Prasad "K-Suryon ,Planck Length and Consciousness Wave Silv Equation of Consciousness
10:Hau,Lene."Physicist Slows Speed of Light" Harvard Gazette.1999

CHAPTER 11
APPENDIX

BRANES;A STRING IS A one-BRANE WHICH COUPLES TO A BACKGROUND SECOND DEGREE TENSOR.Zero-branes are ten dimensional building blocks for space in the pre-Big Bang epoch.The second degree tensor is purported of negligible mass as indicated by the ZERO-BRANE.THE SOURCE OF THE BACKGROUND SECOND DEGREE TENSOR IS R uv where the integral of D to the d power of x where x is the string or one-brane in D dimensions applies to R u v g u v where g u v is the metric acting on R u v for the zero-brane with respect to x which is the one-brane. R u v is the second degree tensor upon which the metric g u v acts. In four dimensions a monopole is dual to two electrons acting on a zero-brane. In 10 dimensions a string is analogous to a five-brane based on p-brane potentials. This involves dual fields such as a tensor R=R* from Ra1...a n=R8b1...b n.p-branes are encircled by a hypersphere which relates to M theory being compactified (curled up)by a circle for typeIIa strings.The charge of a p-brane is based on

$Q= \int \quad * R \, from \, limit \, S \, d - p -$

$2 \, for \, electric \, charges \, and \, Q \int_{Sp+2} R \, for \, electromagnetism..P -$

$branes \, tie \, in \, with \, the \, potential \, involved \, with \, permutations \, of \, a \, field \, tensor. \, p \, brane$
$tensors \, are \, associated$

Are associated with a tensor of the pth rank R a1...a p and electric and magnetic charges can be associated with p-branes with superalgebra.

Dzero-branes represent the vacuum state.ALTHOUGH INDICATED AS ten dimensional building blocks of space they actually are zero-dimensional. ONE-BRANES REPRESENT STRINGS WHICH ARE TWO DIMENSIONAL OR POSSIBLY ONE DIMENSIONAL.If all the dimensions in a system or universe are conserved such that the total number of dimensions are constant;then zero branes would have to be ten dimensional in the vacuum state.Six dimensions for CalabiYau Manifolds and four dimensions of space-time. As the "c" boundary is approached infinite mass with reducing length and width occur when length becoming infinite.In this case width and height approach zero but do not reach it and become infinitely small curled up and compactiifed.In general although showing duality between different systems which are abelian membranes are described by the forces involved with mass or energy associated with the membrane with reference of n-dimensional space where n dimensions would have n-1membranes or n-1 brane.

FLAT OR MINTKOWSKI SPACE IS DESCRIBED MATHEMATICALLY AS THE LINE ELEMENT OR ds2=dx2+dy2+dz2-c2dt2.FLAT SPACE-TIME IS SPACE-TIME WITHOUT ANY CURVATURE IN OTHER WORDS A VACUUM STATE HERE R g a b=0 which indicates that the space-time curvature metric=0 and therefore gravity =0 in the vacuum state.

CURVED SPACE-TIME IS GENERALLY DESCRIBED BY ds2=e-k|r|(dx2+dy2+dz2-c2dt2)+dr2 where r=space-time curvature metric described by tensor as R g a b.R g a b or r is determined by the inertial mass of the object doing the curving and the curving is performed by bosons and possibly gravitons or fermions.Spiral space-time has k=-i(the square root of-1)to the n cotangent theta power as suggested by Dr.Roger Penrose and proposed by this author.

MANIFOLDS
THE SIMPLEST MANIFOLDS ARE CARTESIAN SPACES WHERE A MANIFOLD STRUCTURE OR SURFACE IN TERMS OF TOPOLOGIES IS R to the d power with what's called an identity map Rd implies R d .The coordinate functions of this map are cartesian coordinates.If coordinates are a I ;R d is the manifold of the standard Cartesian coordinates.a i=ax+ay+az and R to the d power is the tolological expression of the standard manifold or the Cartesian Coordinate system. If a manifold is imbedded in another manifold it is a submanifold.On a string basis submanifolds can be orbifolds or Calabi Yau manifolds which are submanifolds for spiral manifolds for asymptotically flat but curved space-time on a macrostatic surface which is expanding and simultaneous rotating as at black hole event horizon.
RIEMANNIAN CURVATURE A RIEMANNIAN SPACE IS THE SPACE COORDINIZED BY xi(power)with a fundamental form of the Riemannian Metric g I jdx I dx j where g=(g ij) obeys the metric tensor.g is of differentiability class C2(all second order partial derivatives of g I j exist and are continuous.g is symmetric g I j=g ji;g is nonsingular |g I j| doesn't equal 0.The differential form and distance from g isn't variant with regard to changes in coordinates.

R I j k l=g l ir(Rr superscript with jkl as a subscriptwhere R jkl with ias a superscript is the Riemann tensor of the second kind.The Riemann Tensor of the first kind is R I j k l=$\frac{\Gamma jki}{xk} - \frac{\partial \Gamma jki}{xi} + \Gamma ilr\Gamma jk$ with r as a superscript $+ \Gamma ilr\Gamma jk$ with r as a superscript $-$ $\Gamma ikr\Gamma ji$ with r as a superscript.**Here**

Γ is a Christoffel symbol or the derivatire of a tensor. Above is Rijkl $- \frac{\partial \Gamma jli}{\partial xk} -$

$\frac{\partial \Gamma jkl}{dxl} + \Gamma ilr\Gamma jk$ with r as superscript $-$
$\Gamma ikr\Gamma jl$ withr as a superscript. Skew Symmetrys involve Bianchi's Identity R ijkl $+$ Riklj $+$ Riljk $= 0$ skew symmetry is R i j k l $= -$Ri j kl an d second skew symmtery is R ijkl $=$ $-$R i j l k with R j k l with i as superscript $=$
$-$R j l k withi as a superscript. Block symmetry is R i j k l $=$
R k l i j. These symmetry properties must fir with the n2(n2 $-$
$\frac{10}{12} COMPONENTS OF THE RIEMANN TENSOR(R i j k l)$
$where the diagnal tensor without s.PTR$

.

Rijkl=g I iRjkl is subscript and I as superscript in the diagonal metric te tensor calculations for the Riemann Metricgives six cases R one R 212 and 1 R 313 and 1 R 323 and 1 R 213 and 1 R 232 and 1 R 123 and 1 which proves with the partial derivatives of the Christoffel symbols of tensors according to the previous formulas give R I j k l=0 for all I j k and l indicatin the summation of all Riemann forces and space is zero.The math of all these combinations is very difficult to reproduce by typing.

GLOSSARY

Abelian:equations having a coefficient or variety in a specific group,g,,algebraic number fields,tensors of the same degree or cohominy group

Anisotropic:not isotropic,lacking observational symmetyry

Anti-symmetric:tensors or vectors that are equal but opposite and can therefore partially cancel or cancel

Aymptotic:that which approaches a level or degree but never reaches it;asymptotic flatness appears without curvature but doesn't reach it

Bianchi's Identity:The identity of groups of Riemannian 4 space that is anti-symmetric and Abelian and cancels each other out of being equal but opposite
"The Big Bang"A theory proposed describing a Friendman type I open expanding f;at universe with is homogeneous and isotropic

"The Big Swirl"A Big Bang with a progressively decreasing rotational vector from an infinite curvature point of space-time to asymptotic flattness

Black Hole:collapsed matter from a neutron star or galaxy with extreme curvature of space-time at the central nexus due to extreme gravity of of the spiral space-time

Calabi Yau Manifold:a surface which represents a relative isotropic portion of space-time with a puckering to accommodate multiple dimensions considered a twisted variant of the orbifold
Choas:absolute disorder

Chiral:a mirror image or absolute symmetry

Closed string:a two or one dimensional building block of matter from energywith movements in 10 or 26 dimensions without breaking the string

Compactified:when every point of the dimensions are curled up mathematically making the size approach zero.First determined by Kaluza and Klein

Conformal Space:when every point in space relative to every other point maintains its relative position regardless of what the space is doing

Dark Matter;an indirectly measured mass causing perturbations in gravity(the curvature of space-time)caused by mass.Acts as cosmic glue containing possibly baryonic particles and neutrinos

Event Horizon:area where a black hole is perceived by measurementsEntropy:degree of disorder

Entropy:degree of disorder
Ex nihilo:out of nothing

M(Membrane)theory:the 5 dual string theories into one massive theory of everything which incorporates membranes which vibrate and incorporate all energy and matter

Isoropic:observational symmetry

Geodesic:a unit of space-time
Gravity:the curvature of space-time caused by mass;actually an effect not a force

Membranes:a description of matter in terms of energy states with stress energy densities described in the number of states with regard to dimensions

N;number of dimensions in N dimensional space

Open string:a two or one dimensional bulding block of matter with movements in a multidimensional plane

Orbifold:space-time manifold in an open twisted cone configuration utilized in string theory

Relativity:the behavior of matter and energy with regard to other matter and energy;energy and space have a different vantage point from other matter and energy including stress energy,time and mass with changes regarding relative velocity

RicciTensor:that tensor which represents inertial mass or resistance against pull or push
Riemann Forces:all strong and weak forces in nature

Riemannian Space:Mintowski space with Riemann curvature of space-time caused by mass.Flat space if no mass is present

Scalar:the magnitude compone t of a vector or tensor with regard to direction

Printed in the United States
By Bookmasters